THE
BLUE
WONDER

Foreword by JILL HEINERTH

FRAUKE BAGUSCHE

THE

BLUE
WONDER

Why the Sea Glows, Fish Sing, and Other Astonishing Insights From the Ocean

TRANSLATED BY JAMIE MCINTOSH

GREYSTONE BOOKS
Vancouver/Berkeley

First published in English by Greystone Books in 2021

Originally published in German as *Das Blaue Wunder: Warum das Meer leuchtet, Fische singen und unsere Beziehung zum Meer so besonders ist* by Frauke Bagusche © 2019 Ludwig Verlag, part of the Random House GmbH publishing group, München
English translation copyright © 2021 by Jamie McIntosh
Foreword copyright © 2021 by Jill Heinerth

21 22 23 24 25   5 4 3 2 1

Greystone Books Ltd.
greystonebooks.com

Cataloguing data available from Library and Archives Canada
ISBN 978-1-77164-604-8 (cloth)
ISBN 978-1-77164-605-5 (epub)

Editing by Tracy Bordian
Proofreading by Alison Strobel
Indexing by Stephen Ullstrom
Front jacket design by Eisele Grafik Design and Nayeli Jimenez
Jacket photographs: (front) © Shutterstock (Rich Carey),
© Getty Images (Darryl Leniuk, nudiblue, Justin Lewis, Mark Tipple);
(back) © Frauke Bagusche
Text design by Nayeli Jimenez

Printed and bound in Canada on ancient-forest-friendly paper by Friesens

Greystone Books gratefully acknowledges the Musqueam, Squamish, and Tsleil-Waututh peoples on whose land our office is located.

Greystone Books thanks the Canada Council for the Arts, the British Columbia Arts Council, the Province of British Columbia through the Book Publishing Tax Credit, and the Government of Canada for supporting our publishing activities.

Canada

For my nephews Miguel and Milo
I hope that you will be able to explore the Blue Wonder
as I have. My greatest wish is that this book
will play a small part in leaving behind a beautiful
and habitable world for you to grow up in.

For Anna, Elias, and Dimi

CONTENTS

~~~~~~~

If there is magic on this planet, it is contained in water.
LOREN EISELEY

*There is nothing wrong with enjoying looking at the surface
of the ocean itself, except that when you finally see
what goes on underwater, you realize that you've been
missing the whole point of the ocean. Staying on the surface
all the time is like going to the circus and staring
at the outside of the tent.*
DAVE BARRY

FOREWORD

~~~~~~~~~~

<span style="font-variant: small-caps">A</span>s <span style="font-variant: small-caps">a child,</span> I was mesmerized by the images of our "big blue marble" relayed from space during the Apollo 17 mission. The cobalt seas, vaporous white clouds, and alabaster ice caps appeared like a boundless wilderness, with little sign of humanity's influence. Astronauts, journalists, and the general public spoke in reverence of the image that changed our perception of our place in the universe. Futurist author Arthur C. Clarke expressed it best, saying, "How inappropriate to call this planet Earth, when clearly it is Ocean." There are few signs of our influence when the sun shines on our blue marble, but we cannot ignore our interconnectedness. The turquoise water that covers nearly three-quarters of the planet's surface appears to flow through everything.

Besides watching the Apollo missions, I was a voracious young reader of non-fiction. *Silent Spring* and *The Sea Around Us*, penned by the American marine biologist Rachel Carson, taught me about science and environmentalism, but also about the wondrous beauty that inspired artists, poets, and visionaries to look to the ocean. Carson wrote, "In nature, nothing exists alone."

Not since Carson's *Silent Spring* have I read a book as motivating as *The Blue Wonder*. Frauke Bagusche's cautionary tale is woven seamlessly with a joyous investigation of the peculiar inhabitants of our liquid planet. Learning that blue whales are louder than jets, that bull sharks swim 2,600 miles up the Amazon to the Andean foothills, or that an octopus may brood for 53 months before her babies hatch enthralled me. Learning that I am more likely to get killed by my neighbor's dog than by a great white shark made me want to call all my friends with unruly pets and scare them in the way that an entire generation was terrified by the movie *Jaws*. Considering that a mother sea otter safeguards her pups by tying them in kelp or that a male lobster protects its partner from predators, one is left to reflect on exactly who the intelligent species is on Earth. Looking at the current trajectory of the Anthropocene, it is perhaps not us humans.

Whether you choose to savor one scrumptious anecdote at a time or dive into the issues and solutions for our rapidly changing planet, *The Blue Wonder* will move you. Enjoy a journey with singing fish, glowing waves, and unimaginable migrations. By discovering the wonders of the ocean, you will be inspired to protect her.

The Earth is 70 percent water. Our bodies are 70 percent water, and we are intertwined in this dance of life. We emerged from the amniotic ocean of our Mother's womb. The sea spills from our tear ducts and sweats from our pores. There is no question that we are water. As we slowly wither with age, and leave the ocean behind, we return to the dust of the cosmos. We will live or die through understanding our blue wonder and the ocean's potential to save us from ourselves.

JILL HEINERTH, author of *Into the Planet: My Life as a Cave Diver*

PREFACE

~~~~~~~~~

I F I HAD to describe myself in a single word, I would say "thalassophile" hits the nail on the head. A thalassophile is someone who prefers living on the coast or by the sea, someone who, quite simply, loves the ocean. The salty, sea-weedy smell, the crashing waves, and the expanse of the sea have an irresistible pull on me. I find it all incredibly relaxing. It is only once I dip beneath the water, however, that I truly experience what I call my "blue wonder." It is there that life is raging. The underwater world has its own tempo and obeys its own rules. It is colorful and in constant motion, and although it is sometimes somewhat uniform, it is always breathtaking.

The mysteries and synergies of the sea have held my interest from my earliest days, so it's hardly surprising that I chose to make my passion my profession. I am a marine biologist, heart and soul, and I have (almost) nothing other than salt water in my head. As a marine biologist the scope of my work is very diverse. Of course, I am not always on the water or under its surface—but as often as possible. Depending on my work, I sometimes don't see the sea for months: say, if I have to analyze samples in the lab, or evaluate data on my computer, or give lectures, or write a book about the sea. When at long last I can once again dive into the sea or snorkel, then I feel as if I have come home.

One of my happiest memories under water is when I was leading a snorkel safari in the Maldives and wanted to show a group of tourists a well-preserved shipwreck. We were just on our way back to the ship when the skipper frantically signaled to look behind me. A few feet away I could see a dorsal fin on the surface of the water. As I couldn't tell with any certainty whether the owner of the fin was a shark or a dolphin, I swam a little closer and saw... nothing at all. A moment later the skipper again began gesticulating wildly and, lo and behold, a pod of dolphins had surfaced behind me. But that wasn't all. All of a sudden bubbles appeared all around us, bubbles from deep down, and we felt as if we were in a whirlpool. Sounds from beneath the surface began to swell to a wild chirping similar to birds ashore. Suddenly we were surrounded by spinner dolphins, chirping and whistling, jumping out of the water next to us and spinning in the air. Beside and under us more and more small groups of dolphins passed by, curiously observing us before launching off on swimming contests. We reckoned that about 300 creatures swam past us, returning to the atoll from hunting on the open sea. Incidents like this leave me speechless (which doesn't happen very often), and I am so thankful that I have been able to experience all these wonderful moments.

Unfortunately, there are also moments that sadden and anger me. I am saddened when an endangered sea turtle dies in my hands because it has become entangled in an old fishing net and I'm not able to free it in time. I am angered when

I walk along a beach and get a backache from collecting all the plastic waste washed ashore. I am angered when I see dying or dead coral because often the harm is human-inflicted and preventable. That said, it is precisely these moments that propel me to educate people about our treatment of the sea.

Although two-thirds of our planet is covered by the ocean and it forms the largest ecosystem on Earth, we currently know only a fraction of what actually happens there. Even the moon's surface has been better researched than the deep sea. The moon and the sea have something in common, however—they both influence us more than we think. The sea we have to thank for our very existence, as every other breath of oxygen we take in is produced by marine microalgae, irrespective of whether you are inhaling air in Denver, Memphis, or Key West. The climate, too, is shaped by sea, from its warm and cold currents, from cloud-producing algae, and from the water cycle. And last but not least, for millennia the sea has offered nourishment and protection, as well as important medicines, employment, and a place to relax. The thunder of its waves, the unmistakable sea breezes, and its great expanse have an enormous appeal for their ability to soothe and inspire.

With this book I would like to share my fascination of the sea with you and to take you on a trip through an environment about which we still know far too little but one that through our daily actions we are continuing to inflict further damage. In a speech to the general assembly of the International Union for the Conservation of Nature in 1968, the Senegalese environmentalist Baba Dioum got to the heart of the matter: "In the end, we will conserve only what we love; we will love only what we understand and we will understand only what we are taught." With this book I would like to awaken in you the love I have for the sea and, with that, the desire to protect this unique habitat. Together we can help the sea regenerate and leave behind a livable world for future generations.

In this spirit, let's dive into the fascinating world of the sea!

THE SECRET
GLOBAL DOMINATION
OF PLANKTON

DON'T WANT TO spoil the fun of your next beach holiday, but when you swim in the sea and swallow some water, you are consuming much more than just salt and water. With every mouthful of seawater, regardless of how clear it seems to be, you are taking the lives of countless viruses, bacteria, certain algae, fish larvae, sea snails (Strombidae), tiny crustaceans, jellyfish (Cnidaria), and arrow worms (Chaetognatha). Your small protein snack is called plankton, from the ancient Greek word *planktos*, meaning "wanderer" or "drifter." All living organisms, whether plant or animal in origin, that float freely in water, are too small or incapable of proper motion, and whose course is determined by the currents are categorized as plankton. Alternatively, creatures that can actively move through water and can also swim against currents, such as fish, squids, whales, or sea turtles, are categorized as nekton, from the same Greek word, meaning "the swimming." Many animal species are both:

they begin their lives as plankton but when mature belong to the nekton group. Organisms that live only part of their life cycle as plankton (usually as larvae) and then, over the course of their development, move through various habitats are called meroplankton. This is in contrast to holoplankton, which have an entirely planktonic life cycle.

Even though most planktonic organisms are absolutely miniscule, they prevail by sheer mass. Nekton account for less than 5 percent of the ocean's biomass; more than 95 percent of marine biomass consists of minute planktonic organisms—the world's secret dominators. In addition to marine environments, plankton is also ever present in streams, rivers, and lakes. Planktonic organisms are split into two main groups: the vegetable phytoplankton, which includes, for example, zooxanthellae, diatoms, and green algae; and the animal zooplankton, which includes larvae, gametes, minute crustaceans like krill, worms, and cnidarians. Plankton groups are also often categorized according to size, ranging from femtoplankton (< 0.2 micrometers) to megaplankton (> 200 millimeters or almost 8 inches), which include species like jellyfish with tentacles a number of feet long.

If you are interested in the diversity of shapes and beauty of this hidden underwater world but have no water samples or microscopes on hand, try to get hold of Christian Sardet's informative photo book Plankton: Wonders of the Drifting World as well as the beautiful illustrations of Ernst Haeckel in Art Forms in Nature. When you have already swallowed countless numbers of these thingies, you should at least go to the trouble of giving the poor unfortunate creatures a face. When I say "countless" organisms perhaps I'm being a little too imprecise. More accurate counts reveal that 4 cups of seawater contain up to 10 billion virus particles, 1 billion bacteria cells, 10 million phytoplankton, and 10,000 zooplankton. Your mouthful of seawater is literally teeming with life. Bon appétit!

Before you pull a face in disgust and swear never to go swimming again—and definitely never ever to swallow even a drop of seawater—take a look in your pantry or medicine cabinet. Those of us who are nutritionally minded don't consume planktonic organisms accidently while swimming; rather, we take them deliberately, namely in the form of pills. Spirulina, a well-known dietary supplement that contains vitamins, antioxidants, and all of the essential amino acids, is extracted from filamentous cyanobacteria of the genus Arthrospira and sold, concentrated, in green pellets.

The use of algae and microalgae by the nutrition and cosmetic industries isn't a modern invention, however. They were valued hundreds of years ago for their positive effects on health, both internal and external. Today, bioactive components such as polysaccharides, chlorophyll, vitamin E, and ectoine are used in skincare products. They help the skin to store moisture, protect against free radicals and UV rays, strengthen the immune system, mask unpleasant smells, and have anti-inflammatory effects. The reason why microalgae produce all these biologically active substances is understandable considering these tiny organisms have to both protect themselves against environmental factors such as UV radiation and develop effective repair mechanisms at the same time. A number of cosmetic manufacturers have even invested in producing their own microalgae so they can harness the biologically active substances for their products.

The medicinal benefits, both for the body and mind, have been long recognized. Research into the use of bacteria found

3

in coral reefs to treat cancer and other diseases has intensified over the last couple of decades (more on this in the "Deep-Sea Drugstores" section). Thalassotherapy—using active agents of seawater, algae, and mud along with exposure to fresh sea breezes and sunshine—has been practiced for centuries. If you take a walk on the beach of Norderney, an island off the northern coast of Germany, a sign praising the health effects of high oxygen, low pollen, and clean seaside air immediately catches the eye. The wind blowing toward the mainland contains rich amounts of aerosols and iodine, which ease bronchial mucus and allows people suffering from asthma or allergies to breathe freely. The bracing effects of sea air also has positive effects on the skin, improves circulation (thus making the body more resistant to stress), eases inflammation, and hastens the healing of wounds. Seaside walks stimulate the metabolism and sleep–wake rhythms become more stable—you simply feel more alive.

The first seaside spa in Germany was located at Heiligendamm resort on the Baltic Sea toward the end of the eighteenth century, and many others followed. Thalassotherapy became less popular in the twentieth century because of its high costs and the advent of new medications. However, even today respiratory ailments, skin disorders, as well as rheumatic illnesses are still treated in thalasso centers.

So, the next time you pop out of the water coughing and spluttering, try to be positive about it. As well as the cosmetic benefits, you have just enjoyed a valuable snack—superfood that cost you nothing. Additionally, and this is far less well known, your next breath is thanks to plankton—or to be more precise, phytoplankton.

Green Lungs

No matter where you are at the moment—Toronto or New Orleans, hiking in Yosemite or relaxing on a beach in Florida— every breath you take is linked to the sea. Phytoplankton, the tiniest of plant organisms (1 to 1,000 micrometers), do in fact produce more than half the global supply of oxygen, which is why they are also referred to as the "green lungs of the sea." Similar to land plants, these minute organisms photosynthesize, using water, carbon dioxide (CO_2), and sunlight energy to produce glucose (sugars) and oxygen as a by-product. The photosynthetically active algae are called primary producers. Sergei Petrovskii, professor of applied mathematics at the University of Leicester, calculated that if the sea temperatures were to rise by about 10.8°F due to climate change, phytoplankton might stop producing oxygen, leading to shortages of oxygen in the atmosphere and ending in mass mortality of humans and animals.

In order to understand what microalgae actually do and why they are indispensable, it's worth taking a closer look at the sea's carbon cycle. In nature, carbon (with the chemical symbol C) is found in its purest form as diamond or graphite. In its compound form it is found almost everywhere, even inside us—or rather, it helps make us what we are. After oxygen (O) with 56.1 percent, 28 percent of the human body consists of carbon (C), a further 14.8 percent is made up of hydrogen (H), nitrogen, calcium, chlorine, and phosphorus, and just a little more than 1 percent consists of potassium, sulfur, sodium, magnesium, and trace elements. But it is not only humans that are formed by carbon: all animal and vegetable biomass consist of stable multiple bonds of these elements, with themselves and other elements. Carbon is, quite simply, the building block of life.

Similar to water, the carbon on our planet is constantly part of a cycle, both above and below water. The atmosphere, the terrestrial biosphere, and water are continuously exchanging carbon. The CO_2 exchange between the atmosphere and the sea takes place in depths of up to 328 feet, the euphotic zone. It comes about due to pressure differences between the atmosphere and the sea. Theoretically, this is a two-way exchange. If the CO_2 pressure of the atmosphere, known as the partial pressure, is too low, then carbon dioxide from the sea is outgassed to the atmosphere. If, on the other hand, the CO_2 pressure is higher, the gas is absorbed in the upper layer of the sea. Nowadays, however, the partial pressure in the atmosphere is permanently higher than in the sea due to anthropogenic CO_2 emissions, and thus is absorbed in the sea.

The total amount of CO_2 absorbed by the sea is 50 times more than the CO_2 content of the atmosphere, and 20 times more than the amounts of CO_2 produced on land by plants and soil. The sea is by far the greatest intake system for CO_2, as carbon dioxide is easily soluble in water—luckily! As long as CO_2 is in the air it doesn't react; it just floats around as a gas, reflects the warmth rising from the Earth, and contributes through the greenhouse effect to the steady warming of the climate. However, when the gas comes into contact with water, it reacts almost completely to other compounds and, in this form, can no longer heat up the climate. For the most part it transforms to inorganic compounds with extra hydrogen or oxygen atoms, such as hydrogen carbonate and carbonate—only a very small part remains as dissolved CO_2. Since the beginning of the Industrial Revolution some 200 years ago, when humans

6

began burning fossil fuels on a large scale and amounts of carbon dioxide rapidly increased, it is estimated that the oceans have absorbed around a quarter of anthropogenic carbon dioxide. However, the capacity of the sea to absorb CO_2 and to transform it is not infinite, and the increasing concentrations of CO_2 in the atmosphere are already leading to serious problems, such as ocean acidification.

In addition to the inorganic carbon compounds, there are also organic ones. Particulate organic carbon is nothing other than a superstructure or biomass of microalgae or primary producers, as the case may be. These microalgae transform CO_2 via photosynthesis to sugar and oxygen while at the same time growing and multiplying then absorbing more CO_2, and so on. If these microalgae die or are eaten by zooplankton and later excreted as fecal matter, they sink as organic particles along with the bonded carbon to the deeper depths of the seafloor and break down into their component parts. This is how CO_2 is sequestered from the atmosphere and transported to deeper ocean layers. This process is called the biological pump. When these organic particles sink and reflect the light, they resemble snow and are thus referred to as "marine snow."

A further transportation of carbon takes place via the so-called physical pump, which, among other things, uncouples from the thermohaline circulation (covered in greater detail in the "(In)Finite Blue" chapter). It's role in the carbon cycle can be quickly outlined: cold water absorbs more CO_2 than warm water; as cold water is heavier, it sinks with the sequestered CO_2, transporting it to deeper ocean layers (downwelling). In the depths it circulates in the slower-moving deep currents before rising, after hundreds of years, to the warmer waters rich in nutrients in the upper layers of the sea (upwelling). There the nutrients are consumed by microalgae for their metabolic processes, and the CO_2 is in part released to the atmosphere.

CO_2 concentrations in the atmosphere are measured in units of parts per million (ppm). One ppm corresponds to one CO_2 molecule per million molecules of dry air. In 2016, global CO_2 concentrations were 400 ppm, while CO_2 concentrations in pre-industrial times were roughly 280 ppm. Without the above-mentioned mechanisms, we would have an even more serious greenhouse gas problem than we already have: the biological pump alone is used so much that, without it, atmospheric CO_2 concentrations would be 150 to 200 ppm above the present figure.

The phytoplankton, which live in the uppermost sunlit layer of the ocean, consume about 108 gigatons of CO_2 annually, which is needed for photosynthesis. That is an incredible amount, and not to be scoffed at when compared to land plants, which consume a total of 123 gigatons of CO_2 annually. The photosynthesizing activities of the many-times-smaller marine algae can be explained by their much faster rate of development (compared to slower-growing terrestrial plants). Phytoplankton can multiply exponentially, and if the conditions for living are optimal (enough light, CO_2, and nutrients such as nitrogen and iron), algal "blooms" can even be seen from space.

Besides these enormous capabilities, microalgae have a larger function relating to climate: they react to weather and can even actively influence it—and, hence, climate, too.

The Smell of the Sea

Beach walks are often characterized by an unmistakable scent—sometimes faint and sometimes intense—reminiscent of salt and seaweed. This typical smell of the sea is caused by the decomposition of algae by bacteria, during which a gas, dimethyl sulfide (DMS), is released. This odor has a higher purpose than solely to exhilarate you while strolling along a beach. Research groups have discovered that when algae suffer from heat stress an increasing amount of DMS is released into the atmosphere: If the algae drifting in the sea become too warm and the UV radiation too intense, they produce dimethylsulfoniopropionate (DMSP) by "sweating" or dying, which is transformed to DMS by bacteria. The DMS rises, is degraded by sunlight to sulfate, and acts as condensation nuclei—particles vital for the formation of clouds. Plainly stated, this means that phytoplankton produce their own protection against the sun by creating clouds (blooms) that, in turn, keep damaging UV radiation away from the algae. Phytoplankton thus play a particular role in influencing the climate, as the radiation emitted by the sun is reflected back into space by the clouds that they create.

The CLAW hypothesis proposes that phytoplankton act as the Earth's thermostat: through the formation of clouds, air, and seawater, temperatures are actively reduced. With the drop in temperature the algae resettle, produce fewer sulfur compounds, and, as a consequence, fewer clouds. In this way, algae are creating their very own feel-good climate. This hypothesis was supported by a study in the South Pacific conducted by the Institute of Marine Sciences in Barcelona in 2011. After evaluating 50,000 measurements from all around the world, the team (led by Aránzazu Lana) discovered that, in the South Pacific, most cloud bacteria originated from single-celled algae. However—and this was the message of the study—global warming could not be regulated by cloud formation of marine algae, as the lessening impact of solar radiation is pretty marginal compared to the amount of greenhouse gases that we produce. Heat stress isn't the only thing that leads to increases in DMS releases. DMS is also used as a protective measure by phytoplankton to ward off predatory attacks by small crustaceans such as krill and other zooplankton. When attacked by krill, algae exude DMS to attract birds that feed off krill. Seabirds like shearwaters, albatrosses, and petrels use the smell of the sulfide compound in searching for food. They can follow the trail over a number of miles, locate the food source, and unintentionally come to the aid of phytoplankton.

Sadly, a bird's fine sense of smell can have fatal consequences, as a research group from the University of California discovered. Maybe you know Chris Jordan's famous photo? It shows a dead Laysan albatross fledgling with a stomach full of plastic. In the photograph you can plainly see a yellow cigarette lighter and a number of plastic bottle caps blocking the decomposed fledgling's digestive tract, leaving no space for natural nourishment and causing it to die of starvation. Unfortunately, birds are not able to distinguish between plastic and food, and this photo is a stark reminder of the hundreds of thousands of birds that perish annually as a result of our dependence on plastic. But what is it that makes plastic so attractive for birds that they confuse it with their natural diet? The research group found out that algae settling on plastic give off higher concentrations of DMS, both to water and air. Attracted by the smell of the gases, seabirds confuse the drifting pieces of plastic with their normal foods. Fish, coral, and other creatures are also enticed by the smell and become victims of plastic—or they pass them on to the next link in the food chain.

Little Snacks for Big Stomachs

Phytoplankton is the basis of all life in the sea as well as the nutritional basis of zooplankton, which include, among others, protozoa, crustaceans, jellyfish, mollusks, worms, eggs, and larval stages of organisms living in the sea. Plankton as a whole—both vegetal and animal—in turn provide food for small and large fish, crustaceans, mussels, coral, and even the planet's largest living creature, the blue whale.

The blue whale, or *Balaenoptera musculus*, belongs to the baleen whale family and can grow up to an impressive 110 feet in length and a maximum recorded weight of 180 tons. This makes them not only the heaviest living creatures but also (quite likely) the largest creatures ever to have lived on our planet, and they are still here. The tongue of an adult blue whale alone weighs roughly 4 tons—about as heavy as an elephant and large enough to hold an entire soccer team. These huge creatures feed off the smallest creatures in the ocean, filtering out krill (tiny shrimp-shaped crustaceans that exist in swarms) and other plankton using their "beards" (baleen hairs attached to the upper jaw). However, in order to feel comfortably full and pleasantly satisfied, a blue whale needs to consume some 15,500 pounds of krill a day.

A wide range of other creatures feed off these small sea dwellers, including three species of sharks: the whale shark

(Rhincodon typus), the basking shark (Cetorhinus maximus), and the megamouth shark (Megachasma pelagios). People who have had the pleasure of swimming with one or more whale sharks will long remember the grayish-blue or brown back with bright dots between light horizontal and vertical stripes, the white underbelly, and the broad head with a bluntly rounded snout. The dotted patterns, covering the back and head, are as unique as fingerprints and help scientists identify individual whale sharks. These gentle giants of the sea can attain lengths of up to 60 feet, making them not only the largest in the shark family but also the largest species of fish in the world. To get a rough impression of their sheer mass, simply stand next to a bus—that's the kind of size ratio to expect when, under water, you find a whale shark snacking on plankton next to you. To reach this imposing size and weight of up to 20 tons, whale sharks, swimming forward with mouths wide open, filter a large number of planktonic and sometimes also nektonic organisms. Every now and then they position themselves horizontally or vertically, and with one huge gulp suck in their prey. Their diet ranges from krill, copepods, and fish larvae to small fish such as anchovies, sardines, and even squids. Frequently, whale sharks are sighted when coral, various types of fish, and Christmas Island red crabs (Gecarcoidea natalis) release gametes—eggs and sperm—into the water, which they then proceed to filter with relish. Whale sharks can be found in almost all tropical and subtropical waters of the world, and—just to name a few examples—can be admired off the coast of the Maldives, Mexico, Australia, and the Azores.

So planktonic organisms are much more than annoying little drifting particles that spoil the view for divers. Without these tiniest of organisms, the air that we breathe would probably run out and, like a line of dominoes once the first one has been toppled, our food chain would collapse. Our existence on Earth would end sooner than later.

Size Isn't Everything

My first experience with zooplankton, on land, was quite early thanks to the contents of *Yps*, a German comic, borrowed from my cousin Marcel. In the 1980s, *Yps*, together with the free prize inside, was the must-have item for children, and the appearance of the latest edition was awaited just as eagerly as the newest version of an iPhone nowadays. One issue came with a little bag of tiny dried eggs from which would hatch shrimps that had existed 100 million years ago. I was totally smitten by the idea of raising my very own brine shrimp (*Artemia salina*) and witnessing their development from eggs to mature reproductive organisms. Although the creatures are pretty robust and not particularly demanding, I still managed to provide the poor things with their very own version of Armageddon and kill them.

Luckily, there are people whose efforts to provide tiny brine shrimp with a long and happy life are far more successful than mine were. The research group of John Dabiri from Stanford University in California discovered something astonishing about these small creatures: For a long time it was thought that zooplankton, due to their size and limited capacity for mobility, had very little or no influence on water circulation in the sea. In a lab experiment in 2018, however, the team demonstrated that these tiny organisms definitely do have a huge effect on churning up the ocean waters when they congregate as a swarm. Brine shrimp were used to represent zooplankton in order to simulate their daily mass migration and to study the mechanics of currents. Using a light source, the phototactic organisms were induced to vertical migration, both up and down, depending on where the light source was placed. This experimental setup reconstructs the natural light conditions of the ocean, where at sunset the creatures swim toward the surface to feed and at dawn dive to the cold deeper layers. In the process, they cover a distance of up to 600 feet—quite substantial when you consider the body size. These diurnal migrations are in some ways comparable to our daily commuter traffic on land, at least as far as biomass is concerned!

Zooplankton make this daily, or rather nightly, journey to forage for food, as their main source, phytoplankton, are found in the upper layers due to their dependence on sunlight for photosynthesis. Staying in light-flooded zones, however, is often risky, as the zooplankton can be quickly found and eaten by other predators. The theory is that in order to minimize the risks, zooplankton seek out their prey at night and hide during the day in the dark depths.

For the first time under lab conditions, the research group was able to demonstrate that a strong down current was generated during the ascent of the tiny creatures and that this motion could mix water masses over a quite a number of feet. This synthesis is beneficial, as the water columns are composed of different layers that are conditioned by the differences in density of the water. So, nutrition can be borne from the lower layers of water to upper layers, which are not so rich in nutrients, with the zooplankton acting together as a kind of marine food blender.

But what do zooplankton do in areas where in winter the sun doesn't shine and there are no day–night cycles? During the polar winters in the North Sea darkness prevails and yet zooplankton still migrate. Until recently it was believed that mass migration of zooplankton just didn't happen in polar latitudes. Kim Last of the Scottish Association for Marine Science in Oban and her team, however, found out that zooplankton in the North Sea migrate just as regularly up and down, but take their bearings from moonlight. When the moon rises, zooplankton drop down to the depths, and when the moon sinks below the horizon, they ascend to the upper layers. During a full moon they sink to even deeper levels to avoid the risks of being discovered by predators.

Glibbery Giants

As explained earlier, zooplankton encompass not only tiny organisms but some larger creatures as well. Some of them have been drifting through the oceans without backbones or brains for more than 500 million years (and I don't mean some human beings—our species hasn't been around that long). I mean medusae—also known as jellyfish. Jellyfish belong to the phylum Cnidaria and are related to coral and anemones. Just like them, jellyfish capture their prey using cnidocytes in their tentacles, which on contact with their prey (or unfortunate humans) inject toxins, paralyzing or killing them. Some resemble floating ghosts in mauve, red, or white; others look like huge flying saucers with thread-like attachments. One thing they all have in common: their fragile, if somewhat slimy, beauty. With perfect symmetry, together with toxins, they wrap their tentacles around inquisitive divers as well as small fish, crustaceans, and other creatures, which are first paralyzed and then digested. Some jellyfish are transparent, making it easy to see what is happening inside them, while others conceal their inner workings beneath a canopy. Jellyfish exist all over the world, and they can be found everywhere from the surface to the deep depths.

The life cycle of jellyfish can be split into two phases: a free-swimming, sexual medusa phase and a sessile asexual polyp phase. The colony-forming polyps, firmly fixed on the seabed, produce free-swimming medusae via budding, which in turn reproduce sexually. After male and female medusae shed their gametes in water, planula larvae develop from the fertile eggs. These sink to the seafloor and begin their sessile life forming asexual medusae—coming full circle.

There are two large groups of jellyfish: Scyphozoa, the "true jellies" (from the Greek for a kind of drinking vessel) and the box jellyfish (class Cubozoa). For both groups the name says it all: the Scyphozoa are cup shaped, and the Cubozoa are more cuboid. The sea wasp (*Chironex fleckeri*), for instance, belong to the latter group. They are sometimes lethal to humans and are one of the most venomous creatures in the world (mostly found in the seas off the northern coast of Australia). Another group, the comb jellies (ctenophora), resemble the "true jellies" but is not classified taxonomically as such, not least because they have no tentacles.

The "true jellies" have radial symmetry, meaning that their body parts are concentrically arranged around an axis. If you rotate them, they look just the same from all sides—there is no back or front—similar to floral blooms cut along the axis of the flower. Humans, on the other hand, have bilateral symmetry: there is a front and a back. One plane divides them into roughly mirror image halves.

The body of a jellyfish consists of 98 to 99 percent water, thus having almost the same density as the water around them. Jellyfish have a flattened, umbrella-shaped bell (mesoglea) on the outer surface where, in some species, you can find simple eyes (ocelli) and the balance sensory receptors (statocysts). The gelatinous bell conceals the inner organs such as the stomach and gonads. From the middle of the bell, starting at the stomach, is the manubrium, a central stalk-like structure ending in a mouth opening. Using another image, the mesoglea with the manubrium look like a mushroom with its stem. In most species, the tentacles trail from the rim of the bell with the cnidocytes, which are responsible for the sting.

The presence of a bilateral nervous system with large accumulations of neurons in the head area gives the impression that jellyfish have no brains and simply drift mindlessly across oceans. But there is at least one scientist who is convinced that moon jellies (Aurelia sp.) are quite capable of directing their swimming movements in response to changing environmental conditions or simply to avoid cliffs or predators. According to David J. Albert of the Roscoe Bay Marine Biology Laboratory in Vancouver, Canada, moon jellies have a central nervous system that acts as a functional, effective brain that can process information and trigger a reaction to this information (a change in swimming direction). Swimming motion is based on the recoil principle in which the jellyfish sucks water into its bell by expansion before expelling it by contraction.

The many types of jellyfish differ both in appearance and size. If you happen to be in Japan and go diving, you might just chance upon a UFO coming in your direction. Nomura's jellyfish (*Nemopilema nomurai*) is the largest jellyfish in the world. It's bell can have a diameter of over 6.5 feet, and this jellyfish has a wet weight of some 450 pounds. Nomura's jellies regularly make the headlines, and this is not only because of their enormous size. These glibbery giants, mostly found in the waters between Japan and China, are famous for their phenomenal propagation. Since the beginning of this century, almost annually, Nomura's jellyfish blooms have posed a threat to Japan's fishing industry, whereas previously blooms were only recorded roughly every 40 years. In summer 2000, the mass of Nomura's jellies was estimated at 94,000 tons (wet weight) on a small strip of Japanese coast a mere 60 miles long. A possible reason for the mass reproduction could be a lack of predators, such as turtles, whose populations have been decimated through bycatch and by the increasing amounts of plastic waste in the oceans. Other natural enemies have probably been overfished, as is the case with the jellyfish's competitors for food. When the competition is missing, then there are more supplies available for the Nomura's jellies, which, unhindered, can grow and multiply. In stark contrast to other marine organisms, jellies can better tolerate, and even benefit from, worsening environmental conditions such as rising sea temperatures and pollution. The financial losses for the fishing industry during such blooms are great, as jellyfish tear the nets due to their size and weight as well as crush fish that have already been caught. Direct contact with the jellies' toxins can lead to severe burning of the skin, which is why crew members have to wear protective clothing.

However, jellyfish do also have their good points, and some of them are fascinating and of great interest to humans. The yearning for immortality is as old as humanity itself—people have searched for millennia for ways to attain it. But it turns out that the key to everlasting life could lie hidden in the sea. Often the great discoveries of scientists happen by accident; take, for example, the discovery of penicillin. And it was pure coincidence that in 1988 a German marine biology student, Christian Sommer, stumbled across immortality during a research field trip on the Italian Riviera. Using a plankton net, he collected Hydrozoa and studied their reproductive behaviors under lab conditions. Among the collected specimens was a small *Turritopsis dohrnii*, now known as the immortal jellyfish. Sommer noticed that when cultivated in a petri dish these tiny cnidarians simply didn't die. On the contrary, their aging cells rejuvenated by transforming back to their earliest developmental stage—polyps. Italian scientists were so struck by Sommer's observations that they took a much closer look at this never previously described phenomenon. In their publication "Reversing the Life Cycle," they concluded that, in all of its developmental stages, T. *dohrnii* could regress to the colony-forming polyp stage, thus becoming theoretically immortal—assuming that the medusa hadn't already been gobbled up by a hungry predator.

In contrast to other documented cnidarian life cycles, rather than die after successful reproduction, immortal jellyfish initiate what is termed a "transdifferentiation program": When a T. *dohrnii* ages or becomes seriously injured, it sinks to the seafloor and folds its bell canopy together. The body degenerates until it looks like a small, slimy blob. After a few days, a polyp develops from it. Once it buds, it produces medusae, which rise like phoenixes from the ashes to swim the oceans as their younger selves. The cycle starts again and, theoretically, could go on forever—were it not for those pesky predators.

After reading these really astonishing scientific findings you might think that the human yearning for imperishability was about to be fulfilled. Sadly, it is and remains nothing more than a dream, as medusae are incredibly difficult to keep in an aquarium and breed. On top of this, far too few specialists are interested in studying these slimy souls. Research funding is granted to supposedly more enthralling or cuddly creatures than these stinging clumps of goo. Luckily, however, a few researchers are preoccupied with jellyfish. One of them is marine biologist Shin Kubota from Japan, who is convinced that the key to eternal youth is indeed in these small medusae. So there is still hope that we can learn something from *Turritopsis dohrnii*.

Jellyfish are not the only ones slinking through the oceans. Other creatures very similar to Scyphozoa are also astir. Take, for instance, the siphonophores: with tentacles that can extend 130 to 160 feet, they are one of the longest animals in the world. Strictly speaking, siphonophores do not exist as single animals but rather in a colony, an amalgamation of many single organisms that together function as one organism. These individual organisms, also called polyps, are linked to one another through ducts, each performing specific functions within the colony. Even if all of the polyps come out of the same embryo, however, they can still look very different from one another. Work in the colony is shared perfectly: while some are responsible for catching prey, ingestion, and digestion (gastrozooids), others take care of buoyancy (pneumatophores), propulsion (nectophores), and reproduction (gonozooids).

Should a siphonophore be attacked and bitten by a predator, no worries: the polyps can rebud and replace the missing pieces.

Of the 175 siphonophores species currently identified, one of the better known is the Portuguese man o' war (*Physalia physalis*), whose tentacles have roughly 1,000 cnidocytes per half inch. When these come into contact with prey, toxins are shot into the unfortunate victim. Paralyzed or dead, victims are then reeled in toward the mouths of the digestive polyps. With the use of its translucent, light blue blending into red "sail"—which is filled with oxygen, nitrogen, and a variable proportion of CO_2—it glides through the oceans. Beneath the up to 1-foot-long, sack-shaped, gas-filled bladder (pneumatophore), which looks a little like half-moon ravioli, are a great many blue, white, and reddish-violet tentacles that extend up to 165 feet in length. The bladder is only activated when there is wind and is kept moist by gentle rocking movements left and right. In the event of surface attack, a siphon structure allows the sails to deflate, enabling the organism to dive.

Creatures like Portuguese men o' war that sail or drift on the surface are technically no longer classified as plankton but pleuston (from the Greek, meaning "to sail or float"). All larger organisms living in the thin surface layer along the air–water interface belong to this group.

During our expedition in 2015, we came across hundreds of these sailing men o' war in the Caribbean. Unfortunately, their numbers were so great that we sometimes couldn't even cast our plankton nets because in no time at all they would fill with a gooey mass riddled with toxic tentacles. Being stung can lead to severe and painful skin rashes. In rare case encounters, it can end in allergic reactions or even death. So, if you see this fascinating creature in water or on the beach, keep your distance and marvel at them from afar, just to be safe.

The Glow of the Sea

Perhaps you have been fortunate enough, during a stroll on the beach or sail through the ocean after dark, to witness the sea glowing. One of the most magical moments I have ever experienced occurred in 2014 on a late-night walk along the beach on the Maldives. The whole tideline—the shore area where all kinds of materials are washed up and remain—glittered in blue and white. It looked as if the whole galaxy had been washed ashore and was illuminating each of my footsteps.

This phenomenon is called bioluminescence or the milky seas effect, and is induced by a chemical reaction that causes a living organism to emit light. In the case of my beach walk, I have to credit my magical moment to unicellular dinoflagellates—phytoplankton with the beautiful name *Noctiluca scintillans*, more commonly known as sea sparkles. *Noctiluca* comes from Latin and means "night lights." The lighting effect is triggered by wave movements and contact between the algae-saturated sands and my foot. This milky sea effect only occurs when concentrations of bioluminescent algae are particularly high—so high that there are algal blooms. You don't have to travel to the Maldives to see luminescent glow; they can be observed pretty much all over the world.

Plankton abundance follows a regular annual cycle and begins with the spring bloom of phytoplankton. When in the early months of the year the days become longer, light supplies also increase and microalgae reproduce phenomenally. As food supplies increase so do the numbers of zooplankton that feed on the phytoplankton, and the seas teem with tiny creatures. When diving or snorkeling, it can, admittedly, be very frustrating to enter the water in joyful anticipation of spectacular sights only to then hardly be able to see your own hand in front of your eyes for all the merrily reproducing plankton.

In the Maldives, I regularly gave lectures on the fascinating world of plankton to ease the frustration of grumbling novice divers who were expecting crystal clear water throughout the year and visibility of at least 160 feet. To prepare for my evening lectures, my team and I would set out in the early morning in our *dhoni*, a traditional Maldivian boat, to take samples with our plankton net. A plankton net is a cone-shaped, fine-grade mesh net with a collecting cylinder at the cod (or tapered) end. Our mesh grade was a mere 0.01 inches. To fish for plankton, we drew the net behind the dhoni very slowly for 30 to 45 minutes. As most plankton gather in the upper layers, you have to take care when collecting samples that the net doesn't sink too deep. To prevent this from happening, a float is sometimes attached to the mouth of the net so it remains stable. Our trawling usually didn't go unnoticed, and we were often accompanied by curious sharks or cheeky dolphins. When fishing for plankton, it is very important to go slowly so that the water flowing through the net isn't so strong that it purées the poor plankton. After successfully trawling enough viewing material for my lecture, we emptied the collecting cylinder into a bucketful of seawater. Because the naked eye cannot recognize anything but a bustling mass, we projected this microcosm onto a large screen with the aid of a camera attached to binoculars. The initial revulsion of my viewers, who had been swimming, diving, or snorkeling in the sea every day, quickly turned to fascination. Thus, I was able to win many new fans for these tiny creatures of the sea. How many people can claim to have seen a torpedo-shaped arrow worm with hook-like grasping spines on its head hunting other zooplankton, or a graceful sea snail dancing through the water, or bioluminescent algae flickering blue and white?

Bioluminescence, by the way, isn't restricted to phytoplankton. It is a common phenomenon among sea creatures. In the course of evolution, a number of species, including fish, echinoderms (sea urchins, starfish, etc.), crustaceans, coleoidea (octopus, cuttlefish, etc.), jellyfish, and bristle worms, developed luminescence independently of one another. Bioluminescent organisms in the sea mostly live in the deep zones or as plankton near the surface. They are seldom found in coastal waters or on the seabed. The glowing effect, depending on the species, is triggered by various reactions and has a variety of functions. In the case of Noctiluca, the effect is triggered by contact and serves as a defense mechanism whereby the sparkle emitted is intended to frighten off a predator. In my case, I have to add, it wasn't particularly effective, but it might be enough to put off a hungry fish larva or convince him to try elsewhere. Another reason for using light is to attract prey—a very popular and widespread method of gaining nourishment in the depths. Admittedly, anglerfish, with their stumpy and sometimes prickly bodies, huge heads, and long fangs with overbite, are not among the best-looking fish, but they have developed a fascinating method to attract prey. Like moths drawn to candles, the females lure their unsuspecting victims with the light organs on their heads before feasting on them. In addition to putting off enemies or searching for nourishment, light is also used to find a suitable partner in the deep depths or as camouflage against other predators. (I go into greater detail about this fascinating phenomenon in the chapter "The Secrets of the Deep.")

CORAL REEFS—
THE NURSERIES
OF THE SEA

WHEN YOU THINK about the largest, most impressive building ever erected, you probably visualize the Burj Khalifa in Dubai (the tallest skyscraper in the world), the Great Wall of China, or the Giza pyramids. However, very few people are aware that one of the greatest living structures was, and still is, made by tiny creatures—coral polyps. The Great Barrier Reef, which consists of almost 3,000 individual reefs off the coast of Queensland in Australia, is an imposing 1,500 miles long and can even be seen from space.

All in all, tropical coral reefs cover roughly 110,000 square miles of the seafloor. This may well be a bit less than 1 percent of the total area, but this almost 1 percent serves as a biosphere for a quarter of fish species found on the planet. Because of this diversity of species and its productivity, coral reefs are often called the "ocean's nurseries" or the "rain forests of the sea." As an ecosystem, coral reefs have the largest density of species, and humans have profited from them for centuries—coral reefs produce the sand of the beaches and sometimes even the islands (atolls) themselves; they offer protection from storm tides and prevent erosion of coastlines; they nourish populations living near the coast; and they are, as has been discovered in the last fifty years, an important source of new pharmaceutical products. Coral reefs are one of the most diverse and most complex ecological systems on Earth, fascinating in their versatility, colorful, lively—simply, a miracle of nature and full of secrets.

But, first, what exactly are corals? Corals are sessile (meaning permanently attached), colony-forming little creatures called polyps that, together with anemones, medusae, and Hydrozoa, are classified as Cnidaria. The polyps most involved with the formation of coral reefs are the stony corals (Scleractinia) (see plates 18–19 in the photo insert), but coralline red algae and sponges also play a role in the formation of reefs. Stony corals, as their name implies, are often mistaken for stone: they can't move and during the day the small polyps mostly retreat into their calcium carbonate skeletons, which resemble stone (see plates 7 and 11, top). At night they venture out of their sanctuaries and capture drifting plankton with their tentacles. If you examine coral polyps more closely, you can see a cup-shaped body with a single central opening that is framed by tentacles.

Reef-forming (hermatypic) corals are found almost exclusively in warm, light-flooded, tropical coastal waters because the special algae with which they have a symbiotic relationship need light for photosynthesis. The distribution zone ranges in latitude from 30° north and 30° south of the equator, and looks like a wide, if somewhat holey, belt that Mother Earth draped around her hips. The majority of coral reefs are in the Indo-Pacific biogeographical region, from the Red Sea to the middle of the Pacific Ocean. Just 8 percent of reefs are in the Caribbean or Atlantic Ocean. Stony and other types of coral can, however, be found in temperate or even cold waters, even in the deep sea. Stony coral, whose skeleton is formed from secreted calcite, belong to the Hexacorallia subclass. Young polyps have six longitudinal walls (called mesenteries), divided symmetrically, that extend from the body wall to the gastrovascular cavity, and each young polyp has six tentacles. As the polyps grow, the number of tentacles and mesenteries increase. The tentacles, armed with cnidocytes, can encircle the opening in a number of rows (the opening serves as both mouth and anus). Besides Hexacorallia, another dominant group of Cnidaria exist in coral reefs, namely the Octocorallia, whose polyps have eight tentacles. This subclass also includes soft corals (Alcyonacea), gorgonians (Gorgonacea), and sea pens (Pennatulacea). Both the hexa- and octocorallians belong to the Anthozoa class, which with some 7,500 species forms the largest group within the Cnidaria phylum. Unlike jellyfish, which are also a subclass of cnidarians, Anthozoa don't progress to a medusa stage but remain polyps. Every polyp in its early stage is there to form a colony, which can consist of many thousands of individual polyps.

If stony corals were to have Facebook accounts, the status would have to read "in a relationship" because most of the species live in symbiosis with unicellular algae, zooxanthellae. In German-speaking countries, "symbiosis" generally means a relationship in which all parties profit. However, internationally (in English-speaking countries) the term "symbiosis" means a general socialization of organisms, without actually

27

having to name the benefits and drawbacks. So, here we have to be a bit more precise: the relationship between stony coral and their zooxanthellae is a mutual symbiosis, as both partners profit from and perfectly complement each other. Zooxanthellae not only live in symbiosis with stony coral but also with many Octocorallia, giant clams (Tridacninae), and even some types of sponge and nudibranchs. In the case of stony corals found in tropical shallow zones, the zooxanthellae live in very high concentrations as endosymbionts actually inside the body tissues of the polyps (at least a million algae cells live in each 0.155 square inch of tissue). These algal lodgers photosynthesize in the safety of the coral polyp's tissues, supplying the polyps with the fruits of their labor—sugar, starch, and other organic products that benefit the growth of their landlord. Without their tiny lodgers corals have a problem, as the shallow tropical and subtropical waters are usually pretty low in nutrients. Corals depend on the help of the photosynthesizing zooxanthellae. The algae, in return, receive security from the corals as well as their by-products, such as nitrates and phosphates, which they desperately need for their metabolism. Most corals living symbiotically with zooxanthellae of the genus *Symbiodinium* have a higher calcification rate than stony corals without such lodgers (azooxanthellate corals).

Calcification is also known as "biomineralization." In biomineralization, a mineral (in this case aragonite, a polymorph of calcium carbonite) is produced by a living organism (the coral polyps). "Polymorphous" describes a substance that can appear in different forms. A biomineral consists of both mineral and organic components, and its production is controlled by organisms. The wonderful thing about biominerals is that they really only reveal their true beauty under an electronic microscope with serious magnification. While working on my thesis, I zoomed in on an oyster shell all the way to the nanometer range and was then able to marvel at the geometric beauty of its individual structures. You can hardly believe that these geometric shapes, which are in no way inferior in precision or variety of forms to geological minerals, are actively created by living animals. Biomineralization is pure art! The process by which the corals form their skeletons has not yet been fully researched, but putting it simply, reef-forming stony corals secrete their calcite skeletons, which are usually made of aragonite, at the base of polyps. Polyps extract the ingredients—calcium and bicarbonate ions—from seawater and in a complex process convert them to aragonite crystals. All this doesn't happen in a jiffy—generally, corals grow only a fraction of an inch a year.

The biomineralization phenomenon is not restricted to corals but extends throughout the animal kingdom. Numerically, the largest group of organisms that form their own calcite skeleton are small unicellular algae called coccolithophores. These tiny, round algae—the most well-known representative is called *Emiliania huxleyi* (also shortened to E. *hux*)—secrete calcium carbonate plates called coccoliths that serve as a kind of armor plating for the algae. The Oceanographic Museum of Monaco has a model of one hanging from the ceiling of the aquarium that looks like a lunar landscape (the design would make a pretty good living room lampshade). E. *hux* is an important primary producer and famous for its extensive blooms, which can be observed from space. Dead algae that sink to the depths play a huge role in the marine carbon cycle, as they transport carbon bonded to their shells to deeper ocean layers (see the "Green Lungs" section).

Other organisms also form skeletons by biomineralization. If your foot has ever had the dubious pleasure of getting to know a sea urchin at close quarters, it surely remembers that sea urchins are very prickly characters. The spikes serve as protection against predators... or careless human feet. The endoskeleton (the skeleton inside the body) and the spikes are both results of calcium carbonate. You can sometimes find these colorful skeletons on beaches after their owners have died. Sea urchins belong to the phylum Echinodermata, which also includes starfish and sea cucumbers.

It is perhaps hardly surprising that we humans also produce biominerals consisting of hydroxylapatite, which is a crystalline calcium phosphate. Hydroxylapatite provides the main component of the inorganic substances in our teeth and bones. Tooth enamel, which has the highest percentage of minerals at roughly 95 percent, is the hardest substance in our bodies. But let's return to the sea. Marine calcified organisms like stony coral, mussels, snails, coccolithophores, or sea urchins are particularly important for researchers studying the effects of global warming on the sea. Numerous studies demonstrate that environments that become acidic can have negative effects on calcification and erode calcified structures.

People who have seen photos of coral reefs off the Maldives, the Red Sea, or the Great Barrier Reef and compared them may well notice the different structures of the reefs. The different types can, in fact, be traced back all the way to the different ways in which the reefs are formed. You can also look at the different reefs using Google Earth. In the Red Sea off the coast of Egypt you will find what are known as "fringing reefs." These are the most common type of reef as well as the youngest. Fringing reefs, as the name implies, fringe the mainland or island coasts and spread out seaward from there. The width of a fringing reef depends on where the seabed begins to steeply slope away from the shoreline; the algal lodgers of stony coral don't receive enough light where the waters are too deep and so coral doesn't grow there. Flattened reef tops below the low tide line are typical of fringing reefs.

31

The most famous reefs in the world are off the coast of Australia—the Great Barrier Reef. Barrier reefs are older reefs and are typically found quite a distance from the coast. Some of them have their origins in fringing reefs. Barrier reefs that were formerly fringing reefs originate through successive eustatic (sea level) changes. The resulting sunken lands form lagoons that separate the barrier reefs from the coast. The word "atoll" is derived from *atholhu*—the only Dhivehi word (the language of the Maldivian population) to have received international recognition. The Maldives are atolls: small circular islands that are made of coral skeletons. As early as 1842, naturalist, geologist, and biologist Charles Darwin formulated the theory that atolls are formed from fringing reefs growing around isolated islands mostly of a volcanic origin, and even today, this theory is still generally accepted. With rising sea levels, the islands sink but the reefs continue growing toward the surface, forming barrier reefs. When the islands disappear completely below the surface, all that remains is a ring of coral with a lagoon forming in the middle and at least one channel to the open sea. An atoll is born. Bores on the Marshall Islands, an island country in the middle of the Pacific Ocean, have discovered coral reef deposits up to 5,000 feet deep, the origins of which go back fifty million years.

Finally, there are the platform reefs or coral cays. These types of reef form independently of land masses and, unlike fringing or barrier reefs with seaward expansion, they grow in all directions. Platform reefs can form anywhere where the seabed is close enough to the surface to provide ideal conditions for coral settlement and growth. Pseudo atolls form if the platform erodes.

As already mentioned, coral reefs are found in both warm waters close to the surface as well as cold, deep waters—some of them almost on our doorstep. On the fringes of the European continental shelf, well covered by the sea, cold-water coral can be found. These jewels of the deep have long been hidden from human eyes, and only in the last thirty years has the sea revealed its deeper secrets. Thanks to developments in technology, teams of scientists, with the assistance of mini-submarines, underwater probes, and other scientific equipment have managed to shed a bit of light on the darkness. We presume that cold-water coral reefs exist in all seas around the world even though the data for these areas is scarcer than it is for warm-water coral reefs. While warm-water reefs are only found in depths up to 330 feet, cold-water reefs range from 130 to 3,300 feet, where the water temperatures are between 39° and 54°F. The complexity of some of these cold coral reefs compares well with their tropical cousins. By means of modern dating techniques, researchers have discovered that some of these reefs are up to 8,000 years old, and the geological data has shown that cold-water corals have existed for millions of years. Cold-water corals consist mostly of stony corals but soft corals (Octocorallia) such as gorgonians and sea fans, black coral (antipatharians), and stylasteridae are also represented. The cold-water coral ecosystems, without light but in seawater relatively rich in nutrients, work very differently to those in shallow water zones. Cold-water corals living in the dark depths have no light-dependent symbionts and have to rely on a continuous supply of nutrients, organic particles, and zooplankton, which is why reef communities are usually found in places with strong currents, such as seamounts. These oases of the deep offer nutrition and protection to many different animal species such as fish, starfish, sponges, and crustaceans, in a habitat that otherwise is rather barren and threatening. The largest cold-water reef discovered so far is the Røst Reef at a depth of 980 to 1,310 feet off the coast of the Lofoten Islands in Norway. It extends for some 25 miles and is roughly 1.9 miles wide.

The fact that cold-water coral reefs are usually found in deep waters unfortunately doesn't protect them from destruction by humans. The development of ever-new and efficient technology now makes fishing at great depths possible and precisely where the reefs are located, the retreat and nursery of a multitude of fish species. Trawling nets leave behind a trail of destruction, and debris is all that remains of a once thriving reef. As corals grow very slowly, it takes decades for a reef to recover. And for things to get even that far, their parents need a lot of luck—and a good sense of timing.

Mass Weddings of the Corals

Coral sex is a tricky business, as corals are sessile creatures who cannot actively decamp and seek out a partner. So, how do you confront the problem of immobility in the search for a suitable partner? That's right, with proper planning! Corals need to plan the sex act meticulously to get it just right (they frown upon the idea of a quickie). Most tropical corals spawn in sync just one or two times a year in a mass wedding on the reef, which is why every year the Great Barrier Reef attracts hundreds of diving or snorkeling sex tourists for this unique spectacle that resembles a snowstorm under water. By night and precisely on cue, the coral polyps release their gametes into the surrounding waters. This synchronous mass spawning of hundreds of different types of corals is triggered by changes in water temperatures, the tides, the position of the sun, and the intensity of the moonlight. The exact timing is difficult to predict, but we do know that the phenomenon only happens at night after a rise in water temperature has stimulated the maturing of the reproductive cells within the polyps. The mass spawning usually occurs shortly after a full moon and lasts a couple of days. It is thought that corals are able to recognize changes in the light conditions by using primitive photoreceptors, although the time of year depends on the location. On the Great Barrier Reef mass spawning happens in the late Australian spring or early summer.

The simultaneous release of eggs and sperm by thousands of corals on the reef increases the chances of successful fertilization and thus the continuation of their own species. Some species are able to stagger the release of their gametes, meaning that, for example, one species starts releasing at 6:30 p.m., another at 7:00, and a third at 9:00. It is thought that by doing so the various coral types are trying to avoid creating a kind of hybridization.

After the release, these reproductive cells, due to their high concentrations of lipids (fats), float to the surface, where the chances of fertilization are greater. If everything goes well, small coral larvae called planula develop, which then search for a suitable place to settle. These wannabe corals are extremely picky about their habitat and can spend days or even weeks drifting through the ocean as zooplankton before finding a site to their taste. Various criteria have to be fulfilled to grow and form a colony: the foundation should be solid (which is why rocks or other corals are well suited), and water temperatures for tropical and subtropical corals should be between 70° and 82°F, a comfortable temperature for corals. Additionally, there should be a continuous current that will provide enough oxygen, zooplankton, and other particles of nutrition. Another very important prerequisite is clear water—they need enough sunlight for the zooxanthellae to perform photosynthesis. Sediment, like drifting sands or mud, settle on corals and can suffocate them so must be avoided. Once all of the right factors come together and the coral larvae have found suitable sites, they then anchor themselves to their base.

Within a few hours of settling, the planula larvae secrete aragonite, sticking themselves to their selected base, and begin the metamorphosis to primary polyps.

In addition to spawning corals, a number of other polyps brood their offspring. Admittedly, this sounds more like chickens than corals, but it's still true. In contrast to the spawning corals that are mostly hermaphrodites, coral brooders are mostly male or female and produce either eggs or sperm in their polyps. The male corals capable of reproduction discharge their sperm in the surrounding waters—the water column—where they find their way to the egg cells of female polyps and fertilize them. Fertilization occurs internally and not in the surrounding water. The fertilized egg cells grow from embryo to larvae and are brooded within the female polyps for a number of weeks. Unlike the spawning corals, which tend to release their gametes into the waters only once a year, brooders are able to produce offspring several times a year. As soon as the larvae are fully developed they are released by their mother polyps. Typically, they are brownish in color due to their zooxanthellae, which are already "on board" even at this stage. These small larvae generally settle relatively quickly to found their own colonies through asexual reproduction.

Most reef-forming types of corals (roughly three-quarters) reproduce, as described, sexually, be it spawning or brooding. But corals wouldn't be so successful if they hadn't other surprises on hand, in this case various forms of asexual reproduction. A widespread variant is budding. In this form of asexual reproduction new clonal polyps are budded from parent polyps capable of reproduction to increase the size of an already established colony or to found new colonies.

Some species like cauliflower coral (*Pocillopora damicornis*) reproduce both sexually and asexually by fragmentation: the coral breaks into pieces, and each of the fragments have the ability to grow and mature into an adult individual that is a clone of the original organism. The breaking apart can happen accidentally—provoked by environmental factors, hungry predators, clumsy humans, or even aquarium enthusiasts—but corals can also fragment deliberately by developing a kind of predetermined breaking point in which a segment breaks away to form a new colony. This fragmentation process ensures that the broken-off pieces of a colony continue to grow and reproduce as long as the fragments settle in a quiet place where it is possible for the polyps to reanchor themselves.

Fragmentation has been used for a number of years by marine biologists for the restoration or rehabilitation of coral reefs. Fragments of a colony are raised in a coral nursery like seedlings in garden centers and then, once they have reached a certain size, are transplanted onto the reefs. In this way, parts of damaged reefs can be restored—a little bit like planting young saplings in stricken forests. While working on a project in Vietnam in August 2018, we made use of fragmentation by selecting chunks of fragmented corals with living polyps. It is easy to find out whether a living polyp is inside a piece of coral, as living polyp tissues are a brownish color while dead tissue is white. We brought the still-living coral fragments to our boat, along with dead coral remains covered in algae that we planned to use as a bedrock, and placed them all in buckets of seawater. Fragments that were a bit bigger were carefully split into 0.5-inch pieces. One by one, we then planted them on individual chunks of bedrock using cyanoacrylate glue and a specially developed catalyst, which ensured that the glue dried quicker and had mild antiseptic properties (see plate 7, bottom). The process had to be carried out quickly in order to keep both the corals and the foundation moist. The use of superglue as an anchoring aid proved a success, and the corals remained undamaged. The lack of harmful effects may sound surprising, but cyanoacrylate glues have been used since the Vietnam War for sealing wounds and is well tolerated by the human body. When sparingly used, the glue dissolves within a couple of weeks. After successfully anchoring the little corals to their bedrock, we took them back to the reef, to a very carefully chosen protected place. We then left them in the coral nursery and regularly monitored them for growth. After a while, the corals begin to secrete calcite, thus securing themselves to the bedrock. The success of a coral nursery can be measured by how well the corals grow and whether other creatures like fish settle there. Once the coral kids grow a couple of inches and are, let's say, teenagers (although biologically they had long been capable of reproduction), they are taken back to the reef proper to continue growing there and to form new colonies. The speed of growth depends on the type of coral and particular environmental conditions.

In Vietnam, Laura Riavitz, an Austrian marine biologist, was our expert for this method, which was developed by the Ocean Quest Global organization and has already been used successfully in a number of Asian countries. In this case, she was showing the process to people working for the Cù Lao Chàm Marine Park off the Chàm Islands. The method is so simple that, in theory, after a thorough training course practically anyone could use it—a glimmer of hope for the threatened coral reefs worldwide.

The Songs of Fish

Of course, coral reefs are home to a diverse community of organisms, many of which have scales and fins—in fact, 25 percent of all species of fish, as previously mentioned. If you have stuck your head under water, whether in the sea, a lake, or a swimming pool, then you are certain to have noticed one or two noises and you might even have wondered where they came from and what caused them. The motor of a passing boat, for instance, sounds as though it is coming from directly above your head even though it is 300 feet away.

The fact that we have so much difficulty discerning the direction and distance of the noise is because water is an excellent conductor of sound waves. Under water, sound waves travel five times faster than through the air. This is particularly true of low-frequency sounds like the deep humming typical of fish communicating (< 1,000 hertz). The speed of sound in 68°F seawater is on average almost 5,000 feet per second, whereas in the air at the same temperature the speed is only 1,125 feet per second. The speed of sound even increases under water with a higher pressure and salinity as well as climbing temperatures.

Unlike us, creatures under water find it easy to determine which directions the sounds are coming from, a necessary skill as visibility is often not so good—it's murky and you can hardly see your own fins right in front of your eyes, not unlike a dense fog on land. On top of this, the light conditions under water become progressively worse the deeper you dive. Knowing where your fellow fish are hiding, or giving out warnings about attackers, could save your scaly skin. The fact that dolphins and whales use sound to communicate, identify obstacles, and search for food may be fairly well known, but less common is the knowledge that other underwater creatures also communicate by producing sounds. It is astonishing just how loud it is under water, especially at bustling coral reefs. People in search of peace and quiet under water are definitely not in the right place.

Contrary to popular belief, fish are anything but silent. Of course, there are a few types of monosyllabic fish, but if you are snorkeling through a reef you can hear a whole choir of fish singing. The lionfish, for instance, lives up to its name by roaring like a lion; batfish (see plate 24), on the other hand, ba-ba-babble to themselves monotonously. Australian scientist Rob McCauley and his team from Curtin University of Technology in Perth demonstrated, with the aid of underwater microphones, that the cacophony of fish can mostly be experienced in the mornings and evenings, rather like the dawn and dusk choruses of birds. While the latter have developed the syrinx for expression, fish have very different mechanisms. Fish can emit a wide variety of sounds: they grunt, croak, roar, click, and hiss. The sounds are used to attract partners, to defend territory from intruders, and to ward off predators.

Two of the main mechanisms of sound production are stimulation of the swim bladder via a variety of specialized sonic muscles, and stridulation or the rubbing together of bones and other hard body parts. The swim bladder is a gas-filled sac that protrudes from the foregut. On bony fish, they can be found in the abdominal cavity and serve primarily to control levels of buoyancy (just FYI, sharks are cartilaginous fish and have no swim bladders). An influx of gas in the bladder raises the buoyancy and expulsion of gas and results in the fish sinking. Gas can be supplied to the bladder in two different ways: some fish gulp air from the water surface and feed it into the swim bladder via a channel in the esophagus; others have a special gland that absorbs blood gases—oxygen, nitrogen, and carbon dioxide—and feeds them into the swim bladder. Some fish, however, have yet another function for their swim bladders, namely sound production. A sonic muscle is attached to the swim bladder that can contract and relax in quick succession and produce a deep drumming sound. The sonic muscles of fish are the quickest contraction muscles of all the known muscles of vertebrates. The uncrowned king percussionist of the fish world must be the oyster toadfish (*Opsanus tau*) who, to attract the girl of his dreams, can perform a pretty fine drum solo with up to 200 contractions per second. The sonic muscles of the toadfish are found on the side of a heart-shaped swim bladder, and when they contract they produce a sound resembling a foghorn.

The second method for producing sounds, stridulation, is made by rubbing together two hard body parts such as teeth, bones, or scales, rather like the chirping of crickets. Some fish use the swim bladder as a sound box. The northern seahorse (*Hippocampus erectus*), for example, produces clicking or snapping sounds by rubbing the edge of its skull against a crown-like bony crest. Among fish, *H. erectus* are unique, loyal creatures that practice monogamy. Once a couple have found each other, they hug and produce a clicking noise. Unlike a number of *Homo sapiens* specimens, northern seahorses devote themselves to reestablishing their bonds by performing a ritual dance every morning. Generally, seahorses are pretty advanced creatures—it is the males that become pregnant. After the females have sprayed eggs into the male's pouch, the eggs are fertilized and incubated there until mini-seahorses develop. During the actual birth, the male anchors himself to a fixed object by his tail and, by bracing himself back and forth, ejects the baby seahorses from the pouch.

One of my favorite fish is the white-spotted boxfish (*Ostracion meleagris*). An adult male is easy to spot among other fish on the reef, as its up to 9-inch body is not streamlined but rather rectangular—yes, like all boxfish (Ostraciidae) it somehow resembles a box. Its head, belly, and sides are a beautiful azure blue, and the sides are dotted with psychedelic yellow blobs. They have a yellow stripe that runs from the head to the tail fin, which encloses a black area with spots of white. With their tiny blue fins, they busily swim through the reef, ferreting out algae, sponges, worms, and other small creatures and eating them. It is hard to believe that the females and young are one and the same species; they look completely different, featuring dark brown to black bodies with lots of tiny white dots and black fins. When it's time to reproduce, the white-spotted boxfish show no mercy. Competing males ram and growl at each other. Also, when spawning they create sounds that could be described as a pure tone with harmonics. I was once growled at by a boxfish when I disturbed his nap. I must have frightened the poor guy to death when I shone my underwater lamp straight at him.

A number of fish species produce noises particularly when spawning, including the striped parrotfish (*Scarus iserti*), which release their gametes among groups of twenty to forty individuals. The whole process looks very funny, as four or five individual fishes break away from the main group in unison, swim from the reef platform toward the water surface like bats out of hell, abruptly turning on their axes while releasing their gametes at the climax of their ascent. Afterward, they dive back down to the reef and safety. The noise they make while spawning comes from the swimming motions and sounds, well, watery. The noises serve, we think, to synchronize spawning and thus to maximize fertilization rates, but doing so has its risks. The sounds and the smell of gametes are irresistible to many predators like sharks, and the poor parrotfish, focusing on their spawning activities, are easy prey.

Although there are a couple of thousand known coral reef fish, comparatively few acoustic types have been described. Researchers Timothy Tricas and Kelly Boyle from the University of Hawaii created a sound library of reef fish with the aid of underwater cameras and microphones as well as rebreathers that mask the interfering Darth Vader–like breathing sounds and the underwater bubbles. The scientists discovered that of the ninety-six varieties of Hawaiian coral reef fish, almost half of them, forty-five, made sounds. They identified a total of eighty-five sounds, meaning that some fish were able to produce a variety of sounds. Almost half of the documented sounds (47 percent) were associated with "agonistic interplay"—behavior patterns connected to rivalry and competition, such as defending feeding grounds or nesting places or chasing off predators. The fish recorded included, among others, representatives of Moorish idols, parrotfish, surgeonfish, butterflyfish, triggerfish, and anemonefish.

But it isn't only coral denizens making noises to communicate. Every year in spring a couple of million Gulf corvina (*Cynoscion othonopterus*) gather to spawn at the northernmost part of the Gulf of California in the Colorado River Delta. During this three- to four-day orgy, which is governed by moon phases and tide cycles, a memorable and very loud spectacle takes place. While spawning, the male corvinas form a choir and, using their swim bladders, create drumming sounds so loud that the oscillations of the sound waves reverberate through the hulls of fiberglass boats and can be heard above water by the naked ear. According to Brad Erisman from the Marine Science Institute in Austin, Texas, and Timothy Rowell from the Scripps Institution of Oceanography in San Diego, California, they are the loudest sounds of sea fish ever recorded. The drumming of the fish can even drown out the noise of outboard motors on panga fishing boats and damage the hearing of other sea creatures. Loud sex, however, has proved costly, as the fish are easy to locate, exacerbating their overfishing and endangered status.

Marine biologist Ben Wilson and his team from the University of British Columbia in Canada studied both the Atlantic and Pacific herring (*Clupea harengus* and *C. pallasii*) for sound emissions and discovered something curious: both species of herring produce sounds by squeezing air from the swim bladder to the anal channel. In our society flatulence is considered impolite, but in the fish world it seems to play an important role. The flatulent herrings, with frequencies from 1.7 to 22 kilohertz covering three octaves, use their skills mostly at night as a means of communication to stick together in a protective shoal. The researchers named the phenomenon Fast Repetitive Tick (FRT) (who says scientific publications are dry and humorless?). The researchers compared the herring farts to a "high-pitched raspberry." And it isn't only sounds that leak out of the herring's anus but at the same time bubbles. Wilson and his team believe that herrings can pick up this frequency while most other fish cannot, so FRT allows the herrings to communicate without alerting predators. There are, however, exceptions, including dolphins and whales, who can hear high frequencies excellently and use FRT as pointers for hunting herrings.

Sadly, humans also produce underwater noises, often at levels that are potentially damaging to animals unable to escape them. All sea creatures, from blue whales to tiny fish, are affected by human-made noises. These sounds range from jet skis to supertankers to underwater seismic trials for oil and gas (which cause explosions that can be heard miles away). All of these sounds produce an acoustic fog that is far above the natural soundscape and makes the communication of different species that little bit more difficult.

Nemo's Brothers

At school most of us learned that males have an X and a Y chromosome, and females have two X chromosomes. The

distinction was that simple. In recent decades we have noticed,
on closer inspection, that things are not quite that straightfor-
ward. There are people who cannot be unequivocally classified
as male or female; they are intersexual. This is also the case in
the animal and plant worlds, but there it is even more complicated. A number of fish species, particularly coral reef fish,
can change sex over the course of their lives. A sex change at
some point in life is called sequential hermaphroditism and
occurs in three main forms. Protogynous hermaphrodites, the
most common form of sequential hermaphroditism, begin life
as females then later transform to male. So, for instance, the
Californian sheephead (*Semicossyphus pulcher*), a type of wrasse,
begins life as female. A variety of environmental factors can
affect its hormonal balance and start its transformation to
male. In the process, the inner organs as well as the color and
size of the animal changes—the smaller females with dull pink
scaly frocks and white underbellies turn into larger males with
black head and tail sections, reddish-orange midriffs, red eyes,
and fleshy forehead bumps. Both sexes have white chins and
large protruding canine teeth, which are perfectly suited to
grinding up hard-shelled creatures. Bluestreak cleaner wrasse
(*Labroides dimidiatus*), native to the coral reefs of the Indo-Pacific,
are another example of protogynous fish. These small fish
live in a harem consisting of one large male and a number of
smaller females. If the male dies, then the largest female begins
to transform to a male, developing male sexual organs within
2 weeks and eventually able to mate with the remaining females. The small blue fish with yellowy heads and a distinctive
black longitudinal stripe that becomes gradually wider as it
nears the tail fin are interesting enough sea creatures anyway—
they maintain cleaning stations—but more about them later.

The second form of sequential hermaphroditism, protandry, involves species beginning life as males and later transforming to females. Definitely one of the best-known fishes and much loved by children, the clownfish or anemonefish (from the subfamily Amphiprioninae) begins life as a male. These animals usually live in small social groups with host anemones (see plate 5, top). A group consists of two adults and between one and several young fish. Generally, the females are larger; the next largest in size are the functional males—the males that are destined to mate and reproduce with the female. The female is at the top of the hierarchy, and this is obvious not only to members of her group but to any diver who dares to venture too close. These small, mostly orange-, white-, and black-striped creatures look cute, but they can become very aggressive. If as a diver you are stupid enough to tease them with your finger, they will snap at it. They once clearly demonstrated to me that I was not welcome near their anemone by ramming my mask and growling at me. If the female dies, then the second-largest member of the group—so the male that had previously mated with the now-dead female—begins the transformation to a reproductive female. The remaining animals in the group then move one step up the ladder, and the next biggest young fish becomes the functional male. People who remember the Disney film *Finding Nemo* may have noticed that the filmmakers failed to ask the opinion of an ichthyologist, a fish expert. Had they done so, Nemo's story would surely have been very different. For those who don't know the film: Nemo is a little clownfish who was born with one fin too small, his lucky fin. After his mother became a snack for a barracuda, his widowed father takes care of his upbringing. Later, Nemo, who grew up as an only child, is kidnapped by divers working

46

for an aquarium business, and his father undertakes an odyssey to find his only son. Of course, father and son return to their anemone after many adventures, and they both live happily ever after. If the story was told from a biologically correct viewpoint, however, it would get an R-18 rating from the censors, if it were allowed to be released at all. The story would go something like this: Nemo's mother is gobbled up, leaving behind, broken and alone, a father and son. The father then begins to transform into a female while simultaneously Nemo develops into a male capable of reproduction. As Nemo is far and wide the only sexually mature male, they mate and produce incestuous offspring. Should Nemo's partner, his former father, die, then Nemo would transform to a female and look for a new partner. Not quite the material that Disney is looking for when releasing animated adventure films for children.

The third form of sequential hermaphroditism involves species being able to transform back and forth between male and female. This type of sex changing is called serial bi-directional hermaphroditism and is found mainly among gobies in the orders Gobiiformes and Paragobiodon. These small reef dwellers live primarily as breeding pairs but sometimes as part of larger groups, among the branches of coral where they can retreat when danger looms. When they live as part of a larger group, only the largest two or three individuals reproduce. When you are as small as gobies and so sparsely distributed throughout the reefs, then, inevitably, you rely on safe hiding places, which can make searching for partners somewhat tragic. Teams of Japanese and Australian scientists have proposed the theory that it could be an advantage to change gender when necessary: when the partner dies or the balance of the sexes is too one-sided, the gobies have a home

47

advantage. They don't have to venture out in search of a partner and can thus avoid the intense pressure of possibly falling prey to predators. Nature is sometimes pretty smart!

Last but not least, in addition to sequential hermaphroditism there is simultaneous hermaphroditism, in which the creatures have both male and female organs at the same time. In the plant world and among flatworms this variation is commonplace, but the practice is less widespread among fish. In fish, these kinds of sex changers can produce eggs and sperm at the same time. The creatures, however, do not self-fertilize (with the exception of the mangrove rivulus, *Kryptolebias marmoratus*) but spawn either as males or females; take, for example, the chalk bass (*Serranus tortugarum*), which can change sex up to twenty times a day. They have developed a very charming reproduction strategy—egg trading. Two partners, who incidentally are also loyal, exchange egg "parcels" that they have laid during a female phase and fertilize them during a male phase. One hopes that the little basses are spared hormonal fluctuations during the female phase, otherwise you can forget all the stuff about domestic bliss.

So, as you can see, there are many forms of sexuality among fish, and one is just as "normal" as the other.

Underwater Hospitals

Life under water is hard, especially if you are ill, with toothache, have an open infected wound, or parasites want to play around in your scaly frock. So, what do you do without the luxury of a friendly doctor just around the corner? This gap in the marine health system is filled by a number of species of animals, including cleaner shrimps (see plate 15, top) and cleaner fish. (By the way, surgeonfish (Acanthuridae) play no role in the reef's health system. Their name is derived from the scalpel-like modified scales on either side of their tail base, which is used for defense.) As the density of practicing cleaner fish is comparable to the number of rural medical doctors— rather scarce—the fish have to know where to find them. A cleaning station could be situated on a prominent coral head or, just as easily, between swaying tussocks of algae. These stations can be compared to a spa or a hospital because it is here that creatures gather for help with a wide variety of problems. Some have bits of their last meal stuck between their teeth; others have wounds with peripheral infections that need treatment. Some have trouble with their scales and need skincare; others have pesky parasites. The clientele seeking help is very varied, stretching from manta rays (see plates 20, top, and 8–9), to sea turtles (see plate 6, top), to tiny fishes. Often the fishy patients have to wait a long time for an appointment. When it is finally their turn they remain motionless, even letting the cleaning station staff into their mouths and gills. It is astonishing that these little cleaner fish have the nerve to swim merrily into the open mouth of a large and potentially dangerous fish. But they know with certainty that nobody is going to have to save their scales, given the relationship they have with each other. This relationship is mutually beneficial: the patients are helped with their problems and the cleaner fish have their next meal—a classic win–win situation. Technically, the term used to describe this relationship is "mutualism," a form of symbiosis that we have already seen practiced by zooxanthellae and corals. Some cleaner fish also have a special tactic to avoid conflict and the danger of

49

becoming someone's supper: they entertain their clients with dance moves. *Labroides dimidiatus*, the bluestreak cleaner wrasse, have developed a vibrating dance that is intended to pacify the hungry client—with success, as studies have demonstrated.

Dance moves are not solely the purview of cleaner fish. Cleaner shrimps of the species *Urocaridella* sp. have taken the art to new levels, developing a rocking dance to peddle their trade. The hungrier they are, the more they go for it, and with obvious reward: during an experiment, client fish, when offered the choice, preferred the wilder rocking cleaner shrimps for the cleaning post over their not-so-hungry brothers, the gentle swayers.

When cleaner fish take on a job, the removal of ectoparasites is only performed under duress. They would much rather eat the scales or, better still, the mucus of their hosts, which unfortunately is very painful for the patients. Bearing this in mind, the cleaners try to seek a balance between duty and pleasure. And the efficiency and speed at which they check out their clients would put any cashier at a supermarket to shame: they can, believe it or not, serve more than two thousand clients a day. The latest data-protection regulations obviously haven't reached these waters yet because during treatment, patient and cleaner fish are not alone—other clients awaiting treatment quietly stay by and watch. The cleaner fish are apparently aware that good service is essential for a successful business and behave accordingly, granting the wishes of their clients. They have learned that in order to gain new customers it is better to be cooperative and eat the unpopular parasites that are plaguing their clients rather than the much more delicious mucus and crispy, fresh scales. An experiment demonstrated that when not observed by other prospective clients, the bluestreak cleaner wrasse preferred clients without ectoparasites over those with them. When, however, there was a witness to their work, the cleaner fish mostly ignored the mucus and eagerly went for the parasites. The research team also observed that potential new customers avoided cleaner fish that nibbled away at the mucus and scales of the client too often and thus caused pain. Customer care comes first for these fish, and the female bluestreak cleaner wrasse that do not abide by the rules are soon made aware of it. As you may recall, this species of cleaner fish live in a harem with a large male and a group of smaller females. If a female bites a client, thus breaking the trust, she is punished by the male. The males even make sure that the punishment fits the crime, meaning the more important (that is, bigger) the harmed client is, the harder the female is punished and harried by the male.

People who doubt whether fish are actually capable of feeling pain or making conscious decisions should read Jonathan Balcombe's book *What a Fish Knows*, and not just because the Dalai Lama recommends it. Balcombe entertainingly reveals that fish have feelings, that they can sense pain and pleasure. While I was working as a diving guide in Egypt I often drove with my guests to my favorite diving place. After we arrived, we would head for the Canyon, the name of an underwater zone. On the way, we would pass a bluestreak cleaner wrasse cleaning station. The cleaner fish were used to divers dropping in and considered us clients. I just had to take my regulator out of my mouth and, in a flash, they were nibbling away at scraps of skin inside my mouth. On occasions when I wasn't fast enough at moving on, the little beasts would swim up to my mask to signal that my time was up, that I was holding up the traffic, and if I wanted to continue making use of their services

51

in the future then I'd better move on. If I were careless enough not to be wearing my diving hood at the station, cleaner fish would be playing around with my ears before I even realized it—a really unpleasant feeling, as they tug away at the tiny hairs on your ears.

And it isn't only divers who are drawn to cleaning stations on the reef. Many fish seem to consciously choose a reef with an efficient healthcare system. Like other ornamental fish, however, cleaner fish are much sought after by the aquarium business. Experiments have shown that the absence of these important fish means that the vitality and health of reefs suffer—when all cleaner fish were removed, the diversity of the reef decreased by 50 percent and fish abundance reduced to a quarter of the population in just 18 months. Here we are only talking about species that are not stationed on the reef but move freely between the individual reefs. The presence of the small cleaner fish not only ensures the diversity of a reef but also helps the learning capacities of its dwellers. People who have tried to do anything intellectually challenging when ill know that the chances of success are generally pretty slim. Only once you have visited your doctor and been prescribed some suitable medication, taken it, and then rested do you feel fit enough to fully concentrate on your work. And it is just the case for our fish friends, as the study group of Alexandra Grutter and Derek Sun from Coral Reef Ecology Laboratory at the University of Queensland in Brisbane discovered. The group has been studying the behavior of cleaner fish on the Great Barrier Reef for more than a decade and have discovered something astonishing: client fish who had ectoparasites removed at a cleaning station showed an improvement in learning capacities. Just like us, illness—in this case the negative impact of parasites—affected the fishes' concentration and their ability to learn something new.

The biology of the cleaner fish really is very fascinating and complex. If you ever have the opportunity to dive or snorkel and observe them at work, you should do so. And, while you are there, you might have the chance to meet a famous patient at the reef hospital—the reef manta ray. In the Maldives, we regularly visited a cleaning station, not least because we knew it was a preferred spot for reef manta rays. Manta rays belong to the genus devilfish (Mobula) and are divided into two species: the reef manta rays (Mobula alfredi), which are found in open waters near the coast, and giant oceanic manta rays (Mobula birostris), which, as the name implies, live farther out to sea. The origin of the name "devilfish" can be traced to the two horn-shaped, rolled-up head fins at either side of its mouth. Funnily enough, apart from the blackish upper bodies and the bat-like wings/fins, mantas have absolutely nothing else devilish about them. These gentle giants of the sea can grow to a width of up to 23 feet—well, at least the giant mantas. The reef mantas can reach about 18 feet and resemble flying carpets elegantly breezing through the oceans. Incidentally, in Spanish la manta means "blanket." The name "manta ray" is derived from its scientific genus name Manta, as it was previously thought that manta rays and mobula rays were two different species. Once genetic research uncovered that it was one species, the genus Manta became the genus Mobula, which is why what once was Manta alfredi became Mobula alfredi—the same applies to the giant manta.

Back to our observations at the cleaning station. Mantas are very conscientious when it comes to personal hygiene and often spend many hours at the cleaning stations letting themselves be thoroughly groomed. If a manta approaches you, take a peek at its white underside. You will notice a pattern of blotches unique to this particular manta (see plate 20, top). For researchers, this blotchy pattern, similar to whale sharks, serves as an identification mark of each individual. Scientists identify the patterns with the help of photos and a special computer program, so they can then assign the animal an ID and follow its migration route once further sightings have been registered. A close-up view of a giant manta (about 6 feet wide) might prove a bit uncomfortable to some water enthusiasts, but I can reassure you, mantas are peace-loving creatures, and the only animals that should be concerned about them are small fish and plankton. Mantas are filter feeders, gathering food by swimming with their mouths open. In the process, they unroll their "devil" horns so the head fins form a funnel, channeling plankton-rich water into their mouths. Manta rays have a tail, but unlike their relatives the stingrays, the poisonous spike is missing. All rays belong to the cartilaginous class of fish, like the shark, to which they are closely related.

When you come across this majestic flying carpet under water, you can extend your experience a little and reduce stress on the animal by following a few behavioral rules: Keep a distance of at least 10 feet, and only approach the creature from the side, as it is easy to frighten it if you come at it from the front or behind. As with all other wild animals, definitely refrain from touching them. Generally, you should be wary of anyone offering diving or snorkeling safaris, and select only those who treat animal welfare as a priority (not the kicks of paying customers). Sadly, far too often, diving guides or other tour guides totally disregard underwater etiquette, allowing their guests to do as they please, which results in stressed mantas, whale sharks, or sea turtles diving to the depths. Even the healing wizardry of cleaner fish and shrimps is impotent against such disorders.

Deep-Sea Drugstores

In addition to being popular places to set up health centers and cleaning stations, coral reefs are also an inexhaustible source for medicaments used by humans. Pharmaceutically active substances against cancer, HIV, and arthritis can all be found there. Of course, the drugs are not simply lying around on the reefs waiting to be used. It is the marine organisms that are the rich repositories of bioactive ingredients used in new pharmaceutical products.

Sea squirts, zoanthids, sponges, snails, and corals all produce plenty of toxins to paralyze prey or use as a defense mechanism. Sponges in particular are considered very promising suppliers of medically interesting substances. In anticipation of possible questions: Yes, sponges are in fact animals, even though, compared to mammals, they are not particularly complex. They are members of the phylum Porifera (meaning "pore bearer") and are multicellular animals that cannot produce true tissues. They consist mostly of two cell layers separated by a thicker, jelly-like matrix (mesohyl). Their shape, color, and size are variable and usually depend on nutrition and habitat. They can spread over a foundation as a thin crust or grow upward and outward as tubes, funnels, or pipes. They can also form tree- or bush-like structures. Their bright coloring is often very striking—they blaze at you in neon yellow, gleam in an opulent red, loom ghostly white, or paint reefs in spots of brilliant orange. Sponges are totally lacking in organs or a nervous system. Water swirls in and out through their many tiny pores, which also assist the intake of food. Most sponges are filter feeders, but pinacocytes consume larger food particles via phagocytosis (engulfing and internal digestion). The majority of sponge species have the ability to form mineral skeletal elements, which serve as both anchoring and defensive aids. These spicules consist of calcium carbonate or silica and are stored in their soft tissues. The natural sponge in your bathroom, on the other hand, lacks these mineral deposits— its tissues are supported by soft modified collagen proteins called spongins.

Sponges can be found in the shallows as well as the deep seas. Being sessile organisms, they are exposed to the attacks of natural enemies without the ability to move. Over the course of evolution, they have developed some effective defense mechanisms to survive becoming a snack in the stomach of a fish or a nudibranch (see plate 11, bottom). They have also had plenty of thinking time, as they have existed for over 600 million years and are thus quite a bit older than dinosaurs and many other primeval forms of life. Admittedly, sponges don't appear to be particularly dangerous—think of Sponge-Bob SquarePants or the squidgy sponge in your bathroom. But still, these organisms are often armed to the teeth (which they don't have) with bioactive substances that often have antibiotic or fungicidal effects that kill bacteria and fungi, as well as a few other defense mechanisms. It is precisely such defense substances that make sponges highly interesting for pharmaceutical research. In the 1950s, antiviral and antileukemic nucleotides were discovered in sponges, providing hope for the battle against cancer. Since then, annually some

800 other compounds have been discovered in marine organisms, half of them in sponges. Substances inhibiting tumor growths and inflammations, substances with antiviral, fungicidal, and bactericidal effects, all make the organisms much sought-after suppliers. And because the research and development of drugs is a time-consuming and costly process, and sponges cannot be removed from the seas in the quantities necessary without endangering both the ecosystem and stocks, scientists are now trying to breed them. Researchers are also working to produce synthetic equivalents. On average, only 1 in over 10,000 analyzed substances extracted from marine organisms ends up on the market.

The poisons in reef dwellers are also of great interest. Cone snails use a neurotoxin, conotoxin, to immobilize their prey. As this slimy reef creature can only move very slowly, Mother Nature came up with something special to help them—an effective hunting strategy. The cone snails (Conidae) live in tropical and subtropical waters, feeding on fish, worms, and other mollusks. Using its radula—a barbed harpoon-like tongue that can grow up to 0.5 inch long—they can inject poison into fishes swimming nearby, which are then paralyzed or die within seconds. With toxins potent enough to instantly immobilize fast-swimming fish, the fish-eating Conus species can also be dangerous to humans. In the worst cases, human contact with the neurotoxin can even have fatal consequences. Unfortunately, these snails are much-loved souvenirs due to the beautiful patterning on their shells. If you are collecting them or tread on one by accident, they may feel threatened and use their weaponry in defense. The toxicity of these small creatures is hard to believe. During my time in the Maldives, a proud snorkeler came up to me one day and showed me a photo, asking me to identify a pretty animal. On display was a beautifully patterned cone snail, sitting comfortably and very much alive on his hand. I was, to be honest, absolutely speechless at his carelessness, which was clearly born of ignorance. When I told him that he was very lucky this creature wasn't frightened into defending itself with its harpoon, he became very green around the gills. He became paler still when I told him about the incredibly fascinating and highly effective toxins of the snail. My plea to everyone: please don't feel that you have to touch everything. Looking is enough—otherwise, it might end in tears...or worse.

But let's turn now to the beneficial effects of conotoxins. The painkiller ziconotide, used to ease severe and chronic pain, is the synthetic form of an omega-conotoxin peptide derived from *Conus magus*, the magical cone snail. Ziconotide, a drug derived from a marine organism, was authorized for use in the United States by the Food and Drug Administration in 2004 under the name Prialt. Approval for the European market followed in 2005.

Dolastatin 10 is another example of a substance derived from a sea snail that is now produced synthetically. This tumor-inhibiting drug is extracted from the wedge sea hare (*Dolabella auricularia*), which can be found in the Indian Ocean. The antineoplastic (tumor-inhibiting) properties of dolastatin 10 was reported in 1972. It was later discovered that the bioactive substance didn't originate from the sea hares but rather from cyanobacteria the snails ingested when they ate. It is pretty amazing that a pharmaceutical—which according to studies has a positive effect against brain, kidney, and colon tumor cells—extracted from a bacterium absorbed during a meal eaten by a snail can be produced in a lab. Of course, this doesn't happen overnight. In fact, it took almost 40 years from the discovery of the substance to successfully produce it synthetically.

Another substance with anticarcinogenic properties (used 59 in the battle against cancer) is found in Caribbean Sea squirts. At first glance these sessile tunicates look like sponges, but they are more closely related to humans. They are, like us, chordates. Chordates possess, at some stage in their life cycle, notochords (dorsal nerve chords), which in the course of the evolution of vertebrates forms a backbone. The antitumor drug trabectedin (also known as ET-743) is extracted from sea squirts or, to be more precise, from mangrove tunicates (*Ecteinascidia turbinata*). The drug is used in the chemotherapy treatment of advanced soft-tissue sarcoma and ovarian cancer.

Bioactive substances can be isolated from bacteria found both in organisms living in the reefs and from reef sediments. Curacin A, a substance used in the treatment of cancer, is based on enzymes extracted from such bacteria. Curacin A is derived from natural products isolated from cyanobacteria *Lyngbya majuscula*, and in laboratory tests has proven effective against the growth of kidney, colon, and breast sarcomas. The Curacin A toxins in the bacteria *Lyngbya majuscula* probably fulfill a defense function against natural enemies. Cyanobacteria are some of the oldest organisms on our planet, and we have much to be thankful for: Roughly 2.3 billion years ago they started producing the oxygen that we so need. This was, as we all know, the building block for all further and considerably more complex creatures.

Marine organisms are a rich resource of active substances for humans, and the possibilities are still slowly being revealed—another reason why it is in our very own interests to sustain and protect these riches, not act like parasites only interested in our own well-being.

Symbiosis, Parasitism, and Other Interactions

The cleaner fish have shown us that, when it comes to the sea, it is often a case of one fin washing another. In their case, both the cleaner fish and their clients shared the benefits of the interplay. When we use the term "symbiosis," it could refer to a number of biological interactions between two different species, including parasitism, commensalism, mutualism, and neutralism. Most people are familiar with "parasitism"—two different species coexist but one of them profits from the relationship while the other suffers. Under "commensalism," one party profits and the other, although it does not suffer, gains nothing. If both parties profit from the interrelationship, then we're speaking of "mutualism." If the coexistence of two species has no effect on the other, then we use the term "neutralism." So much for the theory, however. In real life things are not that simple, as due to a number of stages and transitions in the course of their interactions, it is often not clear what kind of coexistence two organisms have developed. In other words, it's quite similar to our human relationships.

The relationship between pearlfish (Carapidae) and sea cucumbers is a very one-sided affair, a classic example of parasitism. The sea cucumber is a member of the Echinodermata phylum, and its outer form resembles an intestine going for a stroll on the seabed. That is pretty much how they work, too (see plate 16). They ingest food at the front and excrete it behind. The body extends between gullet and anus, which they stretch and contract in order to move. On first or even second glance, most of us would not consider the sea cucumber a beauty, but luckily there are fish for which outer appearances don't matter—what counts for them is on the inside. In the case of pearlfish and sea cucumbers, this is literally true. Some species of Encheliophis (the pearlfish) seek refuge inside unfortunate sea cucumbers. To be more precise, in the rectal cavity. There it is safe and apparently so cozy that the elongated, transparent, and up to 14-inch-long eel-like fish sometimes even settle there in twos. To the detriment of their poor host, they feed on its inner organs and gonads. This relationship is a classic case of parasitism: one organism profits while the partner suffers the disadvantages.

The coexistence between the suckerfish (Echeneidae) and their hosts is another one-sided relationship. The strange-looking suckerfish attach themselves to other fish, usually manta rays, sharks, sea turtles, and marine mammals. Instead of dorsal fins, these fish have evolved a suction pad. Using this suction pad, they create a vacuum and attach themselves to their transporter. Once firmly attached, they can be moved from A to B, often upside down. Suckerfish can live without their hosts, but having a water taxi is an advantage, as they don't have to move themselves and thus save valuable energy. (This energy saved, however, has an equal cost to their hosts, as their drag increases. Also, unfortunately for the hosts, their skin can become irritated or even infected when the suckerfish stick to the same spot for too long.) Additionally, living on a big transporter offers some protection against predators. And sometimes a little tidbit is diverted to the suckerfish.

Some suckerfish also clean their hosts of ectoparasites, which provides a small snack and the only tiny indication of an advantage to the host, who seemingly has nothing else to gain from its clinging companion.

There are quite a number of bizarre interactions among sea dwellers. One of the best-known examples of mutualism is the relationship between anemones and clownfish (*Amphiprion*). The clownfish, colorful and amazing to look at, are unfortunately incredibly bad swimmers and thus easy prey for predators. To combat this, they have developed an interesting survival strategy to avoid ending up a meal in the stomach of a hungry fish: they live among the stinging tentacles of anemones. Anemones belong to the Cnidaria phylum and are related to coral and jellyfish. The astonishing thing about this setup is that the stinging tentacles don't seem to be a problem for clownfish, whereas other fish need to avoid them at all costs. The trick? Clownfish become immune to the anemone's toxins at an early age by repeatedly rubbing their bodies against the tentacles. Little by little, they build up a protective armor plating made of mucus. And how do anemones benefit from all this? Well, the clownfish aggressively frighten off their host's possible predators and are not afraid of larger opponents such as turtles and divers.

Other anemone guests have pincers and heavy armor plating. Shrimps and a number of crab species live, like the clownfish, under the protection of the anemone tentacles. The spotted cleaner shrimp (*Periclimenes yucatanicus*), found in the Caribbean, offers its cleaning services from the protection of its host anemone. By wildly swinging their bodies back and forth and waving their white antennae, they advertise their cleaning station to passing fish. Should a fish be interested in their services, they remain a certain distance from the anemone and wait for the cleaner shrimps to emerge from the protection of the tentacles and remove parasites and dead particles from their skin.

Not all shrimps live in symbiosis with anemones. Pincered creatures also closely associate with fish. Some snapping shrimps (*Alpheus* sp.), also known as pistol shrimps, work together with gobies (Gobiidae). The next time you happen to go diving, keep an eye out for little fish lying on the sandy seabed in front of a hole. These small fish (called watchmen or prawn gobies) are not relaxing; they are on guard duty. Like alert watchdogs, they are on the lookout for possible enemies—and not only for their own sakes but also for the architects of the hole they are guarding. The builder of the burrow is the very hardworking snapping shrimp, who works nonstop on the dwelling, transporting sand through the mineshafts to the outside. The sandy freight is deposited in front of the entrance to the burrow, with the poor-sighted shrimp keeping contact with the goby via its antennae. As soon as a potential danger approaches the burrow, they both disappear lickety-split into the protection of the tunnel system. This association has its advantages: the gobies have a refuge and the snapping shrimps have their very own early warning system.

Speaking of refuges, community gardens have lost some of their garden gnome charm as more and more young city dwellers are getting back to nature to grow their own fruits and vegetables. Our underwater cousins are the same, and some of them are just as busy in their reef gardens as the city dwellers are in their urban gardens. Some species of damselfish (Pomacentridae) cultivate a certain type of filamentous algae. The reason for this lies in their digestive systems. Damselfish cannot grind and digest the cellulose fibers of many species of algae but have no trouble with filamentous algae types. The most cultivated types of algae are filamentous algae from the genus *Polysiphonia*, which, incidentally, are also a popular meal for other fish and easily ousted by other forms of algae. Pomacentridae take great care to look after and protect their gardens so they do not run short of supplies. Should a diver come too close to this garden, the 2-inch-long fish will puff itself up and growl in an attempt to frighten them off. These tiny farmers, found in the sub- and tropical waters of the Atlantic and Indo-Pacific Oceans, are not particularly discerning and can also behave very aggressively toward each other.

The dusky farmerfish (*Stegastes nigricans*) live in tropical reefs where they cultivate regular monocultures of their favorite algae, conscientiously weeding out foreign greenery. Competitors for food, such as the herbivore doctorfish, are mercilessly harried off their land. If algae-eating snails or other invertebrates in the area lose their bearings and stumble across the plantation, vigilant farmerfish will, without further ado, transport them far away from their treasured crop and unceremoniously dump them elsewhere. Both the cultivated algae and the farmerfish profit from the relationship: the algae can grow and reproduce in peace, and so can the farmerfish.

As idyllic as mutual gardening may sound, there are hidden drawbacks for other living creatures. Before cultivating the protected algae garden, the fish have to ensure that they have prepared the best possible conditions, which involves removing the tissues of coral polyps bite by bite. Only then can they resettle their preferred algae on the already cleared land. In this way, with time, the busy gardeners can kill off an entire coral reef in a continuous cycle of eat or be eaten.

PLATE 1

PLATE 2

PLATE 3

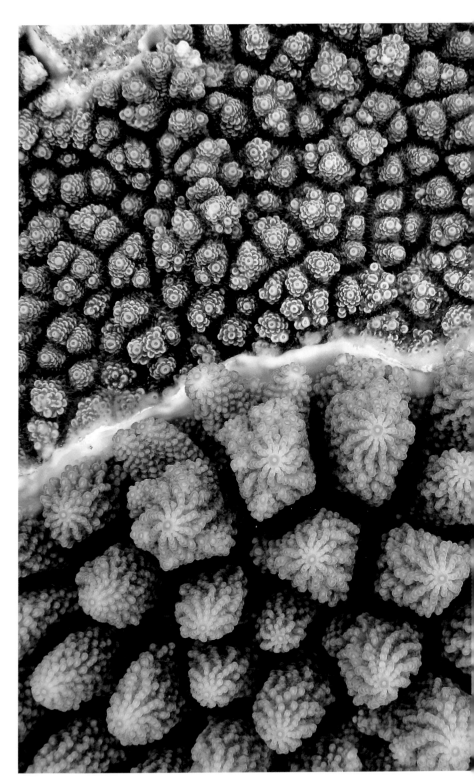

PLATE 4

PLATE 5

PLATE 6

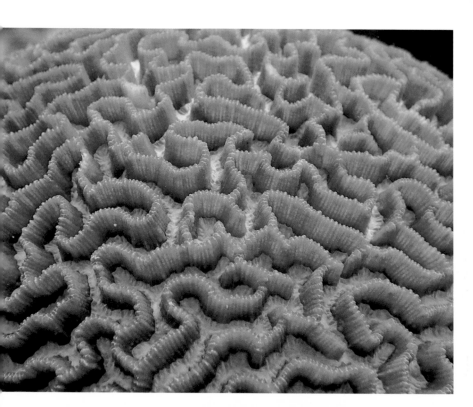

PLATE 7

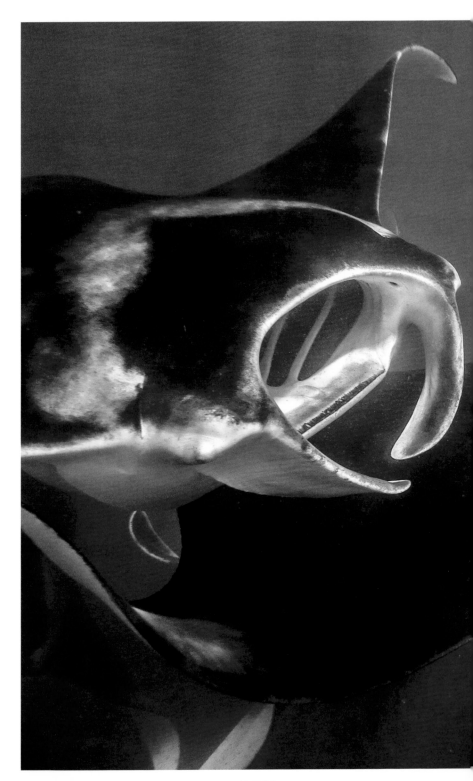

PLATE 8

PLATE 9

PLATE 10

PLATE 11

PLATE 12

PLATE 13

PLATE 14

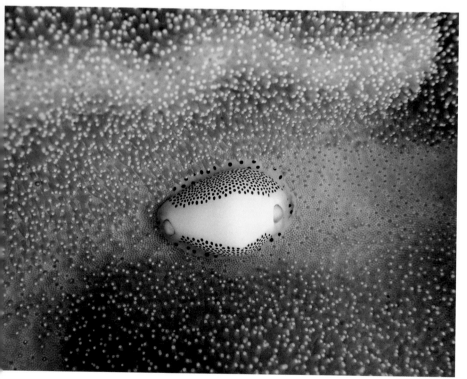

PLATE 15

PLATE 16

PLATE 17

PLATE 18

PLATE 19

PLATE 20

PLATE 21

PLATE 22

PLATE 23

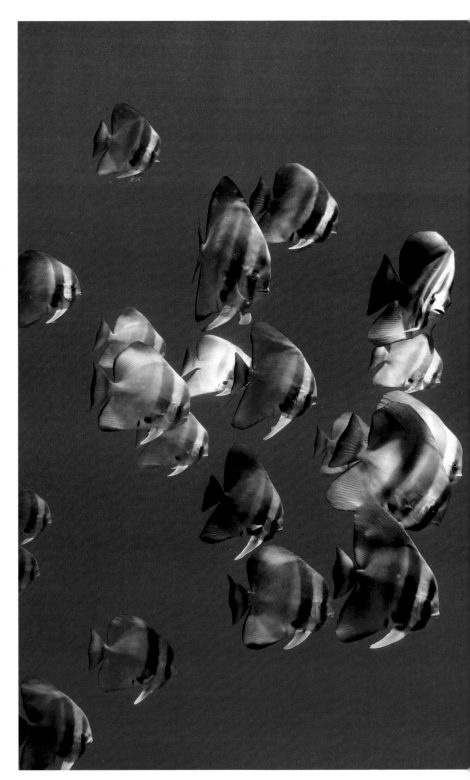

PLATE 24

Coral Reefs as Battle Zones

When you are snorkeling through a coral reef it may well, at first glance, look to be much more peaceful than the hurly-burly world above water. Colorful corals and even more color-ful sponges show off their splendor, algae wave hypnotically to and fro in the gentle current, pretty shoals of fish overtake you, here and there a sleeping turtle (see plate 6, top) hides beneath table coral, and a tiny fish tail sticks out of a crack in the reef where the fish has retired to sleep. But this impression of peace is misleading. In reality you are in the middle of a bat-tle zone in which opponents are eliminated using the meanest methods imaginable. It is nothing less than survival of the fit-test, and if careful observers are lucky, they can witness tactical warfare live. In the Maldives I witnessed quite a few exciting battle scenes while snorkeling. It often seemed as if shoals of reef fish were darting in our direction only to stop and pause beneath us. For me these were always special moments because the fish, of course, were not doing this for fun but rather were using us as protection from their enemies—sharks, mackerels, or tunas. The windier and wavier it was on the surface, the more exciting it was below the waves. There life often seemed to be raging. And the chances of coming across a manta or two with mouths wide open and feeding on plank-ton were fairly good, too. Sometimes you could watch tunas shooting through the water like silver torpedoes, hunting for the next meal. I didn't have the pleasure of snorkeling in rough weather all that often because, for the safety of the guests, the excursions would be cut short or even canceled. On my free days I was able to dive on my own terms and dedicate myself to underwater discoveries. If I was lucky and thoroughly scanned the seafloor, I would be rewarded with an encounter with a very special creature that holds all kinds of records. I'm talking about the undefeated underwater boxing champion of the world who can knock out its prey with one single blow.

The clown mantis shrimp, *Odontodactylus scyllarus* (see plate 5, bottom), is this world-champion fighter, and its KO punch is the fastest in the world. With the power and acceleration of a .22 caliber bullet, the forward movement of its club (technically termed "dactylus"), which has evolved from its second pair of legs, is one of the fastest motions in the whole animal kingdom and will floor its opponent. Within just 2.7 milliseconds, the club accelerates to 75 feet per second, which is the equivalent of just over 50 miles per hour. (To compare: a human blink lasts roughly 100 to 150 milliseconds or 40 times longer.) The movement is so quick that it can't be seen with the naked eye. The clown mantis shrimp gets its name from its resemblance to the praying mantis on land. These insects catch and hold their prey by rapidly shooting out their forelegs and grasping them in a viselike grip. Unlike the claws of the praying mantis, the dactylus is arrested by strongly contractive muscles and a kind of click mechanism on the body. The energy stored while doing this is then suddenly released, similar to how a tension spring works. Hard-shelled prey such as crabs are helplessly smashed and their shells blasted open. As if this wasn't enough, the sudden movement creates vapor-filled bubbles between the club and the prey caused by the large difference in pressure. These so-called cavitation bubbles implode, generating energy in the form of heat and light, and emitting and a loud noise. The resulting shock wave is again at least half as powerful as the initial blow of the club and can even, in some cases, exceed that force. So, if the clubbing doesn't prove enough to sedate or kill the prey, then a combo punch and shock wave should do the trick. These mantis shrimps are categorized as smashers.

A number of species of mantis shrimps have developed a different method of catching prey. Rather than forming a club, their second pair of legs, which functions as their hunting appendage, is covered with spikes and barbed tips. Like the smashers, these shrimps shoot out their forelegs but impale rather than hit their prey. They are categorized as spearers.

Finding mantis shrimps when diving is a particularly rewarding experience because they are not easy to locate. They burrow in soft sediment near coral blocks and are often only visible on closer inspection. Once you have discovered a specimen, however, you are rewarded with a real explosion of color, enough to make circus clowns and harlequins seem drab. Their elongated bodies are greenish to olive brown, dappled like a leopard skin toward the front, and end in a strong fan-shaped tail that facilitates jerky swimming motions. The front pair of legs are greeny-yellow, with brushes on them that act as cleaning aids. Behind those are the second pair of forelegs with the appendages—either the clubs or spears—and another 3 pairs of legs that assist in handling the prey. Between these and the fantail are another 3 pairs of abdominal legs and 5 pairs of swimmers, all orange-red with tinges of mauve. The eyes, situated on stalks, are a kind of shimmering blue. The animal features such a wide range of colors that it can be a bit difficult to describe, but perhaps this is because our eyes, unlike clown mantis shrimp eyes, are really quite primitive. In addition to being boxing champs, these creatures are also super-sighted. Their 12(!) color or photoreceptors can perceive not only polarized and UV light, but up to 100,000 colors. Their compound eyes can move independently and have the greatest number of photoreceptors in the whole animal kingdom. While we have only have to cope with 3 photoreceptors (the cones: blue, green, and red), clown mantis shrimps can pick up colors that would make the wildest LSD trip seem humdrum. This enormous capacity for color recognition probably serves private shrimp-to-shrimp communication, as the shrimps exchange fluorescent and polarized light, for instance, to indicate willingness to mate or when threatening fellow shrimps. The eyes of most clown mantis shrimps are also split into three sections, enabling each eye to have separate spatial views in all directions. You can tell smashers and spearers from each other by looking at the eyes. The eyes of smashers are round, whereas those of spearers are slightly elongated and kidney-shaped. So, you just need to look a mantis shrimp deep in the eyes to know whether it will punch or impale you if annoyed.

When working as a diving guide in Egypt I learned to treasure other highly colorful specimens for their patience being photographed. The roving coral grouper (Plectropomus pessuliferus ssp. marisrubi) is a coral reef and lagoon dweller of the Indo-Pacific Ocean. It has a red or sometimes mauve or blueish body mottled with blue spots and splotches, and a number of dark-colored vertical lines extend from the head along the back. Up to nearly 6 feet long, groupers are much-loved for their beauty (see plate 12, top, of a dusky grouper). Their hunting strategy, which can be observed in the coral reefs of the Red Sea, also makes them fascinating subjects. Together with another redoubtable predator, the giant moray (Gymnothorax javanicus), they team up to form a hunting party. Giant morays can grow to a length of 10 feet and weigh a solid 65 pounds. The nocturnal morays, with their long bendy bodies, can enter small crevices in the reef when searching for prey, whereas the diurnal groupers hunt in open waters. So, the hunting strategies of the two species are very different. While we humans choose people we want to work with according to their suitability, we don't really believe that fish are capable of this. But precisely this kind of behavior has been observed among fish. When a grouper can't get its prey because it has sought security in the reef, the grouper will seek out the help of a giant moray—and not just any old moray but one that has proven cooperative in the past. The grouper will swim to the moray's lair and perform a little dance by waggling its head back and forth.

If the grouper is successful and the moray is willing to wake from its slumber, then the moray will leave its lair and set off on the hunt with the grouper. The grouper will indicate in which crevice its prey is hiding by doing a kind of head-stand and again shaking its head. The moray will then get to work, slithering through the crevice. If the moray catches the fish, it will eat it. If the fish is forced to leave the crevice, then the grouper goes for it. This cooperation between two different species of fish has only recently been scientifically described.

At the access point of a very famous diving spot in Dahab, Egypt, called the Blue Hole, I was regularly greeted by a fish. As a diving guide, I would bring my diving guests here to swim along the beautiful reef slopes. Usually, at a depth of some 60 feet, a long, slender shape, looking somewhat like a swimming stick, would slide into view, seemingly out of nowhere. Swimming slowly, very close to my face, mostly right next to my mask, a bluespotted cornetfish (Fistularia commersonii) would meet my gaze. Once we had briefly acknowledged each other, we would then both get on with what we were supposed to be doing—me diving and it hunting. Each time we met, this swimming "flute" would use me as tactical cover, quickly shooting out from under me to suck small fish and other prey into its long mouth. I can only speculate as to whether it was always the same individual, but it always displayed the same behavior, so I think it was probably the same fish.

Another successful hunter triggers either gasps of rapture or sends shivers down the spine—the reef shark. On Kuramathi, one island of the Maldives, I was able to observe blacktip reef sharks (*Carcharhinus melanopterus*) (see plate 13) from both the jetty and the shore. I didn't experience one single diving or snorkeling excursion without these wonderful creatures. Never before or since have I seen such relaxed and inquisitive sharks, which sometimes swam to within inches of us without displaying any signs of aggression or fear. It wasn't unusual to find a number of them nearby, and sometimes there were up to fifteen reef sharks swimming around me. The largest were about 5 feet long; the smallest just a few inches. The younger animals always kept their distance from the older ones so as not to be eaten. When walking along the beach, we were often parallel to the small sharks minding their own business in the shallows. When swimming, diving, or snorkeling, we were surrounded by the larger specimens. The sharks get their name, by the way, from their first dorsal fins, which look as if they have been dipped in a pot of black paint (see plate 12, bottom). The dorsal fins are often seen slicing through the waters, as blacktip reef sharks prefer shallow waters and swimming near the surface. The gray or sometimes brown base color of their bodies allows them to mesh into their surroundings; their bellies are white. On a number of occasions their excellent camouflage has given me near heart attacks when I would suddenly discover a shark swimming beside me. It is very difficult to describe the feeling of a shark, with its intelligent eyes, sizing you up. I would dearly love to know what goes through their minds. They probably have the very same thoughts.

The encounters with sharks on Kuramathi were definitely something very special, as on the other islands the sharks were rather shy, kept their distance, and swam away when approached. A number of times I witnessed groups of black-tip reef sharks herding shoals of mullet to the shallow zones and then feeding on them. The sharks were also highly active when other fish were busy with mass spawning. Sharks found the smell of gametes irresistible, taking advantage of the fact that the fish were far too busy with their reproduction duties to notice the presence of their natural enemies. Like all sharks, blacktip reef sharks have extraordinary sensory systems, which makes them excellent hunters. First, they have highly developed eyes that enable them to intensify residual light so they can see in almost complete darkness. Their precise sense of smell allows them to sense even the slightest traces of blood in the water, with concentrations as small as 1:10 billion. Additionally, they have a lateral line organ that registers the slightest pressure differences, detecting movement and vibrations in waters surrounding animals. They can also perceive weak electric fields like heartbeats using a specialized organ, the ampullae of Lorenzini. Equipped with this impressive inventory of super sensors, their prey have little chance of escape.

When not distracted by the sight of sharks, you will notice a diverse range of sounds. If you listen carefully, you might be able to distinguish rumbling or scraping noises. Often these are the eating sounds of the parrotfish (see plates 2–3) who, with their parrot-like "beaks," rasp algae from coral, although some species also feed on coral polyps. While feeding off the coral, they leave behind typical chew marks, as they also take part of the calcite skeletons in the process. Their feeding activities not only contribute to bioerosion of the reefs, but, as they excrete sand later, they also help to create sandy beaches. When snorkeling off the Maldives, we swam past an approaching school of garish parrotfish. Once they had passed, it wasn't unusual to swim through whitish-brown clouds of their excretions. Later, I had to explain to disgusted guests that they had spent their whole vacation walking over and reclining for photos on these very same excretions. The grazing fish do indeed contribute to the erosion of the reefs, but they also fulfill an important function—they maintain the natural balance between coral and algae. If the numbers of polyp-feeding parrotfish increase, however, the corals could be so badly afflicted that the natural balance would be disturbed. This often happens when the natural enemies of parrotfish, like sharks and morays, have been overfished.

Another reef predator, the crown-of-thorns starfish (*Acanthaster planci*), is also very fond of reef polyps and can, in a short time, strip a whole reef bare. Its name is no coincidence: with its red body and long poisonous spikes, this species of starfish resembles the crown of thorns given to Jesus. It also looks like a child's painting of the sun with the rays (or in the case of starfish, arms) extending in all directions. Adult starfish reach an imposing size of up to 16 inches, and have between 6 and 23 spiky arms. The starfish inhabit the tropical coral reefs of the Indo-Pacific Ocean. Individually these creatures are incredibly pretty to look at, but if they come together in large groups they can cause such a lot of damage that reefs have great difficulty recovering. Here, too, the increase in populations can be traced back to the elimination of their natural enemies. Sadly, one of them, Triton's trumpet (*Charonia tritonis*), a very beautiful large sea snail, is being fished out of the reefs in huge numbers because the shells of these important predators are popular souvenirs for tourists. Instead of the snails keeping the population of crown-of-thorns starfish at bay, their shells decorate bathrooms worldwide so these nature-loving tourists can dream of tropical beaches the next time they are in the bathtub. Very few people who buy souvenirs from the sea are aware of the effects of their actions—the removal and killing of animals for jewelry, souvenirs, or other novelties leaves huge gaps in the ecosystem.

Other factors also contribute to starfish invasions. The Great Barrier Reef in Australia is frequently afflicted by crown-of-thorn starfish invasions numbering several millions. In addition to the overfishing of the natural enemies of starfish, the effects of agricultural fertilizer washed from fields by rains into rivers and transported by rivers to the sea is another cause. The nitrogen reaches the sea, fertilizes phytoplankton, and causes a rise in the populations of algae. More algae mean more food for the starfish larvae, who develop more quickly and become very hungry starfish. There are plenty of crown-of-thorn starfish larvae, as a single female can produce up to 65 million eggs a year. Often when an army of crown-of-thorn starfish are marching toward hard coral polyps, teams of marine biologists and divers intervene in an attempt at saving the reef from further destruction. There are a number of ways to control starfish; some are injected with poison and removed by hand. People have to be very careful when handling the starfish, as a prick from one of their spikes can cause days of nausea and inflammation, severe pain, and even paralysis. Unfortunately, starfish have insatiable appetites. Adult specimens can consume a plate-size area of polyps per day, and an area of 50 to 150 square feet per year. Once you have a couple of hundred thousand of the species, a coral reef is in deep trouble. On top of this, the multi-armed starfish are no slouches and can munch through a reef at an average rate of about 15 inches a minute. In the 1970s, an eruption in the numbers of crown-of-thorn starfish in the northern area of the Great Barrier Reef lasted all of 8 years. At its zenith, there were some 1,000 starfish per 2.5 acres. Healthy reefs can regenerate after such outbreaks within 10 to 20 years. Sadly, most reefs are no longer healthy but struggling simultaneously with other problems, like the effects of climate change (more on this in the "Endangered Blue Wonder" chapter).

In addition to being threatened by their natural enemies, coral polyps also compete savagely for the best places on the reef. These fights take place silently and slowly, and often go unnoticed by passing snorkelers. The coral reef is no different from the big city: the prime real estate market is fiercely contested and competition for the best places is stiff. Sponges have developed a slick strategy to gain the edge on the reef. They are masters of chemical warfare and resolutely overrun entire reef colonies. Sea sponges like *Terpios hoshinota* in the Indian Ocean coat coral reefs like crusts. If the fish that keep the sponge populations in check are removed from the reef, then the coral, of course, are at a competitive disadvantage. Then there are also fewer and fewer sea turtles who, among other things, feed on sponges. Sea turtles are hunted and killed for human consumption, land as bycatch in fishing nets, and their eggs are considered a delicacy. Once their predators are missing, sponges can gain the upper hand on the reef. This has already happened in various reefs in the Caribbean. However, I still can't help admiring sponges. In Guadeloupe, a group of islands in the Caribbean, I was able to snorkel through whole forests of sea sponges—huge Day-Glo yellow sponges that resembled funnels or tubes growing in crystal clear waters between soft coral (see plate 17). Simply spectacular.

Not only do corals battle sponges for the best spots, but they also battle other corals—and, like sponges, they too use chemical agents. It is estimated that in one reef alone between 22 and 38 percent of coral are at war or encouraging their neighboring polyps to fight, to keep their space, or to prevent it from being overgrown (see plate 4). One of the strategies they use to fight might strike us as being a bit peculiar: corals "vomit" juices containing cnidoblasts (stinging cells) over their enemy that "digest" them. In another method, they use specialized tentacles (longer than their normal ones) as sniffer dogs, feeling out the neighborhood for intruders. These tentacles retract on contact, but before doing so release a load of cnidoblasts that harm the invaders. Species of the genera *Euphyllia* and *Galaxea* are well known for their sweeper tentacles, which look a little like spiders' thread swaying in the air and are many times longer than normal tentacles. All of the tentacles are usually loaded with the cnidoblasts, which are fired off during close-quarter interactions. These perforate the tissues of invaders, releasing their toxins, which normally leads to the death of the affected area. Another tactic often employed by corals of the same species is mucus secretion. The mucus containing toxin or cnidoblasts coats and harms the enemy. The last but not least effective method of securing a good place on the reef is simply overpowering neighboring polyps that don't develop as quickly. If you are observant you can watch corals fighting each other, but there usually isn't enough compressed air in the tank to discover the outcome of the attack—this war takes place in slow motion.

Luminous Corals

We have already established the cleverness of corals but that they are able to glow is genuinely sensational. Unlike glow-worms and other bioluminescent animals such as the deep-sea anglerfish that use chemical reactions to light up (more about bioluminescence in the "Glitter in the Dark" section), stony coral polyps create light by absorbing sunlight and later releasing it. This process is known as fluorescence.

When sunlight hits water, some of it is reflected, some of it absorbed, and some disperses. With reflection, the light hitting the water surface is refracted and the water reflects, for instance, the color of the sky. Incoming longwave radiation is absorbed in the top layers, while shortwave radiation, in clear waters, penetrates to depths of 328 to 492 feet. The long wavelengths of the spectrum—infrared, red, orange, and yellow—can penetrate depths of 32, 49, 98, and 164 feet, respectively. Short wavelengths—green, violet, and blue—reach right down to the lowest levels of the euphotic zone. Blue penetrates the farthest, which is why deep, clear seawaters and some tropical waters seem to be blue colored most of the time. Additionally, clear waters have fewer particles to influence the light transmission. The dispersion through the water itself also affects the color. Waters in shallow coastal areas tend to have larger numbers of particles, which disperse or absorb the light wavelengths differently, so waters near the coast seem to be greener or browner on account of the sediment particles.

The fluorescing objects absorb light of certain wavelengths and then emit them. In the process some energy is lost in the form of heat, so the outgoing light has less energy and a longer wavelength than the incoming light. The change in wavelength means a change in color. Many stony coral polyps found in tropical waters have special fluorescing pigments in their tissues that absorb light and emit it later as pink, violet, green, or yellow light. We can only make informed guesses as to why they do this. Fluorescent corals seem to be more resistant to coral bleaching. The higher the density of fluorescent pigments, the greater the resistance (more about coral bleaching in the "Endangered Blue Wonder" chapter). The fluorescing pigments are photoproteins, and the most recent theory is that these act as a kind of sunscreen, preventing too much UV radiation from damaging zooxanthellae. The photoproteins cover the zooxanthellae in tissue, protecting them from damaging wavelengths that they later emit as light.

Since at depths of more than 49 feet corals don't need this protection from the sun (because the UV radiation is absorbed by water), it was thought that deeper corals wouldn't fluoresce. So it was all the more astonishing to discover that even at depths of 164 feet some corals radiate vivid colors. Researchers puzzled about this phenomenon for a long time, as it wasn't clear what the whys and wherefores of luminescence actually were. Jörg Wiedemann, a coral expert at the National Oceanography Centre in Southampton, solved the puzzle. Together with his team, he discovered that corals produce their own light to encourage their symbiotic algae to photosynthesize.

The algal lodgers of stony corals, the zooxanthellae of the genus *Symbiodinium*, need, like all plants, sunlight to photosynthesize. At depths of more than 328 feet only the blue parts of sunlight—those with wavelengths of around 470 nanometers and the greatest range—can penetrate. Researchers discovered that, in this case, the photoproteins of the fluorescing corals lay beneath the zooxanthellae, where they absorbed blue light and emitted orange-red or greenish light—light with longer wavelengths. This adjustment meant that corals settling at deeper depths could still supply the light necessary for the photosynthesis of their algal symbionts. A win–win situation.

You need special equipment to observe the colorful spectacle of corals when diving. The best way to marvel at the glowing of corals is in darkness. So, in addition to a normal underwater lamp to light the way, you need a special UV underwater lamp and a yellow filter for your mask. When you shine the UV light at the coral, you suddenly feel as if you have landed in a surreal painting or a rave party in the 1980s because the coral reef gleams in all its glory—neon-colored corals of green, mauve, yellow, orange, and, from depths of 65 feet, red. The colors only become visible to the human eye with the yellow filter on the mask, which is like putting glasses in front of the mask. The yellow filter means that UV light from the lamp is filtered out, and we can see the colors that the fish can. If you don't use the filter, then everything lit up by the UV lamp will appear blue. Many reef fish have natural yellow filters built into their lenses. Fish are tetrachromatic organisms (from the Greek *tetra* = four, and *chroma* = color), with an additional UV light receptor on top of the three color receptors of red, blue, and green. We humans, on the other hand, are a trichromatic species and, as we discovered when covering the clown mantis shrimps, have only three color receptors. So, to marvel at glowing corals we need an artificial color receptor—the UV lamp—together with the yellow filter on the mask. The filter has to be yellow, as this is the complementary color to blue or ultraviolet light. Nowadays, diving centers increasingly offer "fluodives," night dives with UV lamps, providing the necessary equipment. People who have reservations about night dives can place red filters on their masks and at a depth of 49 feet admire red fluorescing corals with the naked eye. The red range of sunlight is completely absorbed, and the red fluorescent corals clearly stick out against the background.

79

Perfectly Hidden

How often during a dive do you see someone pointing excitedly at something so you swim closer and see... nothing? You look again and still see nothing. Of course, it is quite possible that it is high time to make an appointment with your eye specialist. It is far more likely, however, that you fail to see the animal because of its camouflage. Some camouflage experts like the angler frogfish (*Antennarius* sp.) (see plate 10) blend into the background and are very difficult to distinguish from their surroundings. If they are really good at hiding, then they are less likely to have trouble with their enemies. If you really strain your eyes, you can sometimes recognize a slight movement, an irregularity in the full picture and then... click... you see the outline of the creature. This has often happened to me, and every time I'm always really excited about unmasking the camouflage artist.

It is quite likely that most of us have already made the acquaintance, either in water or grilled on a plate, of one of the most successful masters of the art of camouflage in the whole animal kingdom—the octopus (see plate 20, bottom). Eight arms equipped with suckers, three hearts, eyes as highly developed as ours, a completely boneless body, a neural network that extends throughout the body, and the capacity to make itself invisible all make this creature fascinating. Octopuses belong to the class Cephalopoda (from the Greek words for "head-feet"). The name fits perfectly, as the "legs" are attached to a sack-shaped body that looks like a huge "head." This head is the mantle that houses the internal organs. The eyes are positioned left and right on the head, between the mantle and arms. The arms are grouped around the mouth, which is fitted with a parrot-like beak. Octopuses also are jet propelled, via a siphon. This tubular organ is responsible for their locomotion through water, among other things. Water is pumped from the mantle cavity through the siphon to the sea outside, propelling the octopus in the opposite direction. The siphon is also used for respiration as well as the discharge of waste products and ink when threatened. It is thanks to special skin cells called chromatophores that the animals are able to blend with the background in a fraction of a second by adapting to the color and structure of the surroundings. These cells near the skin surface consist of various colored pigments and are elastic. By contracting muscles, the animals can actively transform the size of these cells, resulting in color and pattern changes. Octopuses can even change the surface structures of their skin, with their repertoire ranging from warty to smooth and all the stages in between. The color change helps with hide-and-seek as well as communication and chest-beating. If two rival males meet, a color duel ensues. Usually, the male that conjures up the darker to almost black shade wins.

The gold medalist of the underwater camouflage contest has to be the mimic octopus, *Thaumoctopus mimicus*. This species, first discovered in the 1990s, is found, if you can see it at all, in tropical waters. They are anything but easy to see, first because these creatures are mostly found in estuaries, where it is usually muddy and visibility is bad, and second, because T. *mimicus* imitate other species with deceptive accuracy. And when I say other species, I don't mean a couple or a few. I'm talking over 15 documented imitations of diverse animals, ranging from sea snakes, lionfish, jellies, flatfish, or crabs, depending on the situation. Scientists refer to the ability to master camouflage as "mimicry."

The ability to change color isn't confined to real octopuses, the ones with eight limbs. Their near relatives, squids and cuttlefish, are also cephalopods but have an additional two arms. Cuttlefish live near the seafloor and are no less masters of camouflage than the octopus. Depending on the surface they are hiding on, they can look like sand or stones. There are even cuttlefish who can deceive their own species by disguising themselves as females. When a male mourning cuttlefish (*Sepia plangon*) woos a female, he acts strategically and cunningly. Using his pigment cells, he produces a manly pattern on his body that resembles zebra stripes to win over her three hearts. The pattern, however, doesn't appear all over his body but only the part of the body facing the female. On the other side of his body he displays the brown-white speckled markings typical of females. Should another male rival be in the vicinity, he will think that there are just two females, hanging out together. This gives the clever male more time to convince the female that he is an ideal partner and mate before the rival can drive him off. This behavior is not only smart but also points at highly developed mental capacity. It's difficult to believe that these creatures are related to snails and mussels, isn't it? Like them, cephalopods belong to the Mollusca phylum. But true to the first part of their name, "head-feet" have the largest brains of all invertebrates.

The next member of the camouflage guild is somewhat less charismatic than octopuses and cuttlefish, and always appears to be in a bad mood. We are talking about the reef stonefish (*Synanceia verrucosa*). With its ungainly body and warty-looking skin, its large, sullen, downturned mouth, and its huge head, it is not a pretty sight. However, these animals are really fascinating. Their stumpy algae-covered bodies blend in perfectly with their surroundings. You only happen upon this solitary seafloor dweller with a lot of luck, as you can hardly tell them from solid rock. Stonefish imitate structures of their habitat with their inherent form and color, which they cannot actively change. Technically this tactic of resembling something not of interest to the observer is called mimesis. Their camouflage is mostly beneficial in ambushing prey. As they are bad swimmers, they spend most of their time idling on the seafloor. If a small fish, crab, or octopus passes by they open their huge mouths and suck in their prey. Their unspectacular exterior is deceptive, as these fish are some of the most poisonous fish in the ocean. The poison contained in the spikes of their dorsal fins is one of the most dangerous animal toxins and can even prove lethal to humans. Humans, however, very rarely come into contact with stonefish, and then only need to worry when they accidentally tread on one. Sometimes the fish can be found in shallower waters of the Indo-Pacific Ocean and the Red Sea, and are disturbed by unsuspecting waders. But, I should mention, the fish are not aggressive, and the neurotoxins are usually only released for defense purposes or against predators, although to someone writhing in agony after being stung, this is probably totally irrelevant.

Less harmful, at least for humans, are the pygmy seahorses (Hippocampus bargibanti). The name alone induces a "Gee, how cute!" and a sigh. And that's just how it is. These tiny seahorses, who don't grow to more than an inch, live in groups among gorgonian sea fans (Muricella paraplectana), which they clasp onto with their tails so as not to drift off with the current. There are two variations in coloring: their little bodies are either light gray or yellow and are covered with red or orange tubercles. Tubercles resemble warts but despite this still look cute on these creatures. (My dear stonefish, it's a cruel world!) With these adornments pygmy seahorses resemble the gorgonian sea fans to such an extent that it is very difficult to tell them apart from the branches of the coral. For underwater photographers the ultimate goal is first to spot these creatures, at least once, and then to photograph them.

Ghost pipefish (Solenostomidae), which are related to the Syngnathidae family of fish that includes seahorses, pipefish, and seadragons, are other beautiful reef dwellers that also blend perfectly with their habitat. Ghost pipefish look very pretty but, depending on the type, also very bizarre. Some can pass for seagrass leaves; others seem to have fallen off a cactus, as their patterned body is covered with prickles. The coloring and skin appendages suit their respective environments.

Coral hinds (Cephalopholis miniata) are distant relatives of the Red Sea groupers. Their whole bodies are speckled with small blue spots. The tail, anal, and dorsal fins are bluey black, the pectoral and pelvic fins orange or orangey red. With such extravagant patterning you would think that in a reef they stick out like a sore thumb, but far from it. It is precisely this pattern that allows them to blend in with the reef. The color red is the best camouflage in water—what seems to really stand out on land doesn't appear red a few feet below the surface but rather a brownish color on account of the absorption of longwave radiation. The Red Sea groupers prefer to linger beneath rock spurs or coral ledges during the day, places where little light reaches, which makes the fish appear brown. Their spots make the silhouettes very difficult to distinguish from the dim background so that they blend into the surroundings. This form of camouflage, whereby the animal optically "melts" into its setting via patterns on its body or coloring, is called visual crypsis. Many reef fish use this form of camouflage, which accounts for their complexions.

Cute seahorses, charismatic octopuses, grumpy old stonefish…all of these reef dwellers use camouflage. But so do most sea creatures. Maybe not quite as sophisticated but still effective is the glistening blue/silver colors of many fish in the open waters. Very often their underbellies are white in contrast to their upper bodies. If a predator approaches from below, the fish, with their whiteish bellies, are not easily visible. A predator approaching from above looks down into an increasingly darkening ocean, where blue/silver silhouettes blend well. This type of camouflage is based on the principles of countershading, sometimes called Thayer's Law. The body side facing the sun is darker, and the side facing the dark, paler. In this way the fish are almost invisible. Other species take invisibility literally. Many jellies, some fish, and all zooplankton are colorless, and their bodies are translucent. With translucent bodies they don't stand out at all against the surroundings, the infinite blue of the open ocean.

(IN)FINITE BLUE

~~~~~~~~

A T FIRST GLANCE the open ocean, the pelagic zone, looks like
a desert—but with water. Monotonous, an endless
expanse and with hardly anywhere to hide, it must be
hard for sea creatures to survive there. Well, that is the conventional wisdom. The truth is far from it! The open ocean,
stretching from the surface to the deepest depths and with
water masses continually in motion, abounds with an enormous variety of life optimally adapted to the living conditions
there. Speedy hunters, communication experts, and marathon
swimmers all allow themselves to be carried with the currents
and swim through the different zones of the water column.

Many organisms of the open oceans live their existences
without ever coming into contact with shores, the seabed, or
the surface. They spend their whole lives surrounded by water.
This is barely imaginable for terrestrials like us, who breathe air,
enjoy the sun, and like to feel the grass beneath our feet.

The pelagic zone is by far the largest habitat on our planet,
and in order to better understand this huge ecosystem, various scientific disciplines have divided it into different sections,
each of which offers a unique diversity of organisms. Plankton,

as we have already mentioned, is found in the sunlit pelagic layer. This upper zone, called the epipelagic, stretches from the surface to about 650 feet. Here, biodiversity is at its greatest, and apart from plankton all sorts of fish are bustling about, including sharks, mammals like whales and dolphins, sea turtles, crustaceans, and octopuses. Everything found below 650 feet is considered the deep sea. From 650 feet to 3,300 feet is the twilight zone, the mesopelagic. The lack of light means that no phytoplankton exist here but instead bizarre-looking creatures like the bristlemouths (Gonostomatidae), many of whom have green- or red-glimmering light-producing glandular organs called photophores. Quite a bit of glowing occurs in this zone, and most of the bioluminescent species are found here. Below 3,300 feet things get pitch black, but here, too, the zones are differentiated. Depths down to 13,000 feet encompass the bathypelagic zone, followed by the abyssopelagic, which stretches down to a depth of some 20,000 feet. Lastly, the hadopelagic zone descends to the deepest parts of the ocean at roughly 36,000 feet.

Many species spend most of their lives in one zone or the other, but some species do cross the boundaries, such as sperm whales (Physeter macrocephalus) and leatherback sea turtles (Dermochelys coriacea) who can dive to depths of 3,300 feet when hunting prey.

Water—A Substance With Special Characteristics

Water, or H_2O, is a most extraordinary molecule, as it is the only substance that can exist in our atmosphere in three different states—liquid, solid, or as a gas. The two hydrogen (H) atoms asymmetrically bind with one oxygen (O) atom to produce a water molecule. While the oxygen atom is negatively charged, the two hydrogen atoms are positively charged, making the whole molecule a dipole. Like little magnets, the positively or negatively charged poles of every water molecule each turn toward the oppositely charged pole of the neighboring molecule, forming a loose bond. If it becomes colder, like humans, the molecules huddle together—density increases, volume decreases. If it becomes warmer, they distance themselves from one another—density decreases and the bonding becomes increasingly loose until finally the water moves to a gaseous state in which the molecules float around independently. Water molecules are at their densest at 39.2°F, meaning that at this temperature the water has its greatest density and smallest volume. If it gets even colder and water turns to ice, the density decreases instead of getting denser. This phenomenon is called the water anomaly. The result of this anomaly is that ice always floats on the water surface, whether in a rain puddle or the Atlantic, which is just as well for fish, who don't want to be crushed by sinking sheets of ice, as it is for us—if tons of Arctic ice were to sink to the ocean floor, sea levels would be considerably higher. At the point of change in physical state the molecules react relatively sluggishly, needing temperatures of 32°F for fresh water to solidify and 212°F to get it to boil. Salt water behaves slightly differently because of the higher salt content. The Mediterranean, for instance, with an average salinity of 3.74 percent only freezes at 28.56°F.

It isn't, therefore, just temperature that influences the density and freezing point of water; salt content also plays a role. In a nutshell, the saltier the water the lower the freezing point and the higher the density. The sea as a whole has an average salinity of 3.47 percent and is, at the same temperature, roughly 3 percent denser than fresh water. The higher density is also the reason that when you jump into salt water you don't sink as far as you would in fresh water, and why it is easier to float on the surface. There are, however, great differences—in the Dead Sea you can almost lie on the surface without dipping into the water, whereas in the Baltic Sea you have to paddle furiously to remain afloat. And greater buoyancy in salt water isn't reserved for humans—ships also float higher, meaning that they can carry more freight in salt water versus fresh water.

The influence of temperature and salt on the density of water plays a central role in the pelagic zones. As cold water is heavier than hot water, and salty water is heavier than less salty water, cold salty water sinks. Via a process known as convection, water masses circulate. Incidentally, you can recreate the process in a simple experiment in the kitchen. All you need is a glass, some warm water and some cold water, and a little food coloring in order to see the convection a bit better and make it look a bit prettier. This convection of water masses is the most important factor in the creation of ocean currents. But let's discuss salt a little longer.

Why the Baltic Sea Tastes Less Salty Than the Mediterranean

Some people may have wondered why the sea tastes more salty and your eyes burn more in the Mediterranean than in the Baltic Sea. The answer is complex, and I will have to expand a bit. The surface of our planet consists of over 70 percent salt or brackish water. This has a total volume of almost 340 million cubic miles and is just a bit more than 97 percent of all water on Earth. The rest is fresh water from streams, rivers, lakes, glaciers, and groundwater. Less than 1 percent of this almost 3 percent fresh water can be used by us as drinking water. The Zentrum für Allgemeine Wissenschaftliche Weiterbildung (Center for General Scientific Continuing Education) in Ulm in southern Germany came up with an interesting analogy: if the total global water supplies were to equal a full bathtub (about 300 U.S. liquid pints), then the fresh water proportion would fit in half a bucket and the amount of drinkable water for humans would fit into a shot glass. This water, the aggregate of salt and fresh water, has existed for roughly 4.5 billion years (since the birth of our planet), its state of matter continually changing in the individual spheres. The water cycle is maintained mainly by two powers: gravity and the sun. In a simplified overview, the seawater evaporates reaching the atmosphere as moisture; this condenses in the atmosphere and precipitates, falling back to the Earth's surface in the form of rain, hail, or snow. In constant cycles of evaporation, condensation, freezing, and thawing, the water changes its physical state of matter from liquid to solid or gas and back to liquid.

The answer to the question of why the Baltic tastes less salty than the Mediterranean or the Atlantic lies in the water cycle. As rainwater has low levels of carbon dioxide (CO_2) and is thus slightly acidic, when it hits the ground and rocks it can dissolve the mineral salts stored there, which are then transported to the sea via rivers. In salt water, the dissolved salts consist of over 90 percent chloride ion and sodium. When seawater evaporates, sodium chloride (NaCl), also known as the cooking salt that we like to use for meals, remains. The levels of salinity vary from sea to sea and, as mentioned, are on average around 3.47 percent of the total mass. The differences in salinity are predominantly related to the amount of rainfall, the number of river influxes, and the degree of evaporation. The Baltic tastes less salty because the number of rivers flowing into it is considerably greater than the number flowing into the Mediterranean, and as it is farther north, there is more rainfall than in the subtropical Mediterranean. On top of this, on account of the higher temperatures in the Mediterranean, the evaporation rates there are quite a lot higher than in the Baltic. Just to give a few examples: the average salinity of the Baltic is 0.8 percent, the North Sea 3.5 percent, the Atlantic 3.54 percent, the Red Sea 4 percent. The saltiest waters of the world can be found not in a sea but in Ethiopia: the Gaet'ale Pond has a salt content of 43 percent, followed by the Don Juan Pond in the Antarctic, which has a salt content of 33.8 percent and, even at a crisp -104°F, refuses to freeze. The third place for saltiness belongs to Lake Assal (Djibouti), a crater lake in East Africa. The Dead Sea comes next in the rankings, which contrary to its name is not a sea but a lake with a salinity of 28 percent. But not all salts deposited in the seas come from land. Hydrothermal springs and underwater volcano eruptions, where salts from the outflowing lava are dissolved in seawater, also provide a steady flow of mineral salts.

Maybe it is the salt content or mineral salt content that links us to the sea. I often feel that I have salt water instead of blood gushing through my veins, which according to a theory of the evolutionary biologist Martin Neukamm may contain a grain of truth. In his book *Darwin heute* (Darwin

Today), Neukamm compared the mineral contents of human blood and seawater, and discovered that the ratio of the biologically important ions sodium, potassium, calcium, and chloride in our blood plasma is 94:3:2:70. This means that for every 94 sodium ions there are 3 potassium, 2 calcium, and 70 chloride ions. The substance ratios of these ions in seawater is almost identical with 94:2:2:100—proof, according to Neukamm, of our watery ancestry. Whether we can believe the theory or not remains to be seen. For my part, I think it is certainly exciting and plausible, and if it were really true it would explain my deep association with the sea.

The Ocean as Climate Kitchen and the Marine Conveyor Belt

There is no such thing as small talk without touching on the weather. In my mind, the Brits are the uncrowned kings of weather small talk. I have never spent as much time chatting about the local weather conditions as during my time in England. It was impossible to escape any situation without exchanging a few words about the rain, the sun, and the rain again, and any attempt to do so was immediately considered treasonable. If you happen to be traveling to Great Britain in the future, then gather information about local weather daily, preferably a number of times, so you will be on top of any small talk!

But how does weather come about, what does climate encompass, and what role does the sea play? First and foremost, it is very important to understand the difference between weather and climate. Weather is a short-term state of the atmosphere at a particular place and a particular time. Weather is experienced and felt personally and directly. When you are taking a walk and it begins to rain, then you feel the rain. Luckily, rain can quickly change, and the shower of rain in which you were just standing turns into beautiful, warming, and drying sunshine. Climate, on the other hand, is defined as a long-term state—the mean readings of weather over a number of decades but at least three. In the process all sorts of weather parameters are measured over a long period of time and statistically analyzed. So, climate cannot directly be felt.

The role of the sea is decisive for climate, and here we come back to the water cycle and convection. Differences in the density of water, induced by diverse temperatures and salinity, and assisted by winds, are the driving forces behind the world-wide convection known as thermohaline circulation (THC) or, sometimes, the ocean conveyor belt. (The term "thermohaline" has Greek origins, *thermos* = warm, hot; *halas* = salt.) Primarily, this circulation is initiated in winter via the sinking of heavy, high-density salt water to depths of around 6,500 feet. This sinking, the convection of water, is confined to the convection zones of the poles—in the North Atlantic, the Greenland and Labrador Seas, and in the Antarctic, the Weddell and Ross Seas. Putting it somewhat simply, the cold-water masses, rich in salt, flow as deep-water currents toward the equator, where they become warmer and lighter and rise. The warm waters then flow as surface waters north, evaporating and cooling then sinking again, forming a completed cycle connecting three oceans (the Atlantic, Pacific, and Indian).

The water masses of this marine conveyor belt are also affected, although to a considerably lesser extent, by winds and Coriolis forces. The Coriolis force is named after the

Frenchman Gaspard-Gustave de Coriolis who in 1835 published a paper on his mathematical evaluations. It is defined
as the inertial or fictitious force of the Earth's rotation. Pole-
ward linear movements, like currents, are diverted toward
the east (to the right) in the northern hemisphere and to
the west in the southern hemisphere. The total amount of
water involved in the cycle amounts to 96,000 cubic miles
or roughly one-third of all the oceans' water. Surface currents
transport enormous amounts of heat. One significant surface
current and, in Europe anyway, the best-known section of THC
is the Gulf Stream. It contributes to the moderate climate of
Western and Northern Europe compared to other areas on the
same latitudes. Without the Gulf Stream transporting warmth,
it would be on average 8° to 15°F colder in this region. The Gulf
Stream System is a fast-flowing surface current in the Atlantic
Ocean with average water temperatures of 79°F. It transports
more water from the Caribbean to Northern Europe than the
total sum of all the rivers in the world and in its warmest spots
can reach temperatures of 86°F. The system is supplied by the
warm-water masses of the north and south equatorial currents.
Beginning in the Atlantic off the west coast of Africa, the current, assisted by trade winds, heads for the Caribbean, where
it reaches its highest temperature, then through the Gulf of
Mexico, through the Straits of Florida as the Florida Current
with a mean transport of over 1 billion cubic feet per second.
Once the Florida Current joins the Antilles Current, the current becomes what we know as the Gulf Stream, which flows
north along the eastern coast of the United States. Near Cape
Hatteras in North Carolina, the Gulf Stream veers off toward
the northeast due to the Coriolis force and the prevailing west
winds. Here it encounters the cold Labrador current flowing

south from the Arctic Ocean and joins a branch of it to become the North Atlantic Current (NAC), with a marked loss of heat and power. In the east Atlantic the current forks. The Canary Current heads south along the west coast of Africa and joins the North Equatorial Current. The first cycle is complete. The other fork continues as the NAC to the coast of Ireland, past the west coast of Scotland to the North Sea. There the water sinks to the depths, as it has become much colder and more salty, and flows back into the Atlantic as deep-sea current. The second cycle is also complete. This ever-moving convection, and with it the transport of heat, is the reason that people in the southwest of Ireland can stroll beneath palm trees.

Unfortunately, according to two studies published in the science journal *Nature*, Europe's heater in the North Atlantic is showing signs of weakening. The Atlantic circulation (scientifically known as the Atlantic Meridional Overturning Circulation or AMOC) is losing power and has been for quite a while. Levke Caesar of the Potsdam Institute for Climate Impact Research identified two "fingerprints" that were characteristic of the weakening circulation: a cooling down of the sub-polar Atlantic Ocean and a heating up of the Gulf Stream region. David Thornalley and colleagues at University College London concluded that the circulating currents have become 15 to 20 percent weaker in the last 150 years. If the circulation continues to weaken, then impact on the climate in North America and Europe could be dramatic, with alterations in air temperature and greater risks of storms triggered by transformations in the route of low-pressure areas caused by changes in sea-surface temperature.

The Gulf Stream, however, does more than distribute enormous amounts of heat. It is also used by creatures great and small as a means of transport to get from A to B without too much effort. The reason the Gulf Stream is such an attractive method of transport for sea creatures such as loggerhead sea turtles (*Caretta caretta*) and humpback whales (*Megaptera novaeangliae*) lies not only in its speedy flow but also in its onboard service with comprehensive buffet. The nutrients carried out to sea by the east winds from the Sahara are particularly popular with phytoplankton, the staple diet of zooplankton. Zooplankton include larval stages of the widest variety of organisms and other small crustaceans that are eaten by small fish, which are the food of larger fish and other animals, and so on throughout the marine food chain. So, the welfare of all passengers is well catered.

The Journey of Sea Turtles

The sea turtle is one of those animals that causes the heart of all sea lovers to skip a beat. Anyone who has seen a turtle asleep beneath a table reef or watched one busily chomping knows what I mean. When the animal then eyes you curiously, without shyness, it really is one of the most beautiful experiences in the world. There are two families of sea turtles: the Cheloniidae and Dermochelyidae. The Cheloniidae includes six species: green sea turtles (*Chelonia mydas*), olive ridley sea turtles (*Lepidochelys olivacea*), loggerhead sea turtles (*Caretta caretta*), hawksbill sea turtles (*Eretmochelys imbricata*), flatback sea turtles (*Natator depressus*), and Kemp's ridley sea turtles (*Lepidochelys kempii*). Only leatherback turtles (*Dermochelys coriacea*) belong to the Dermochelyidae family. All sea turtles, unlike their terrestrial relatives, have streamlined bodies with flattened carapace. Their extremities have adapted to their marine habitat and become strong flippers. The salt glands behind their eyes are a further adaptation to their environment. These glands constantly release a concentrated salt solution, thereby helping the kidneys regulate the salt content of the blood.

The leatherback sea turtles, unlike members of the Chelonidae family, have no bony shell. (The hard shell of the hawksbill in particular sadly is used far too often for producing tortoiseshell.) Instead of the solid carapace, leatherbacks have a softer shell made of thick leathery skin with embedded miniscule bony plates. Adult animals can have a curved carapace length of up to 6.5 feet and weigh on average 650 to 1,100 pounds, but animals that weigh up to 1,500 pounds have been found. This makes the leatherbacks the largest of all sea turtles. Their large teardrop-shaped bodies are the most streamlined of all sea turtles, and their front flippers, which can be 8.8 feet long, are the longest of all swimming marine reptiles. The body of the leatherback is blue black or dark gray, with white or pink spots. On the back, seven distinct ridges run from the neck to the tail. In addition to their enormous size and prehistoric appearance, they have an impressive ability to dive really deep. Animals usually dive in the epipelagic zone, and rarely deeper than 820 feet, but individual leatherbacks have been recorded at depths of 4,000 feet, making them some of the deepest-diving creatures in the world. As if this weren't enough, leatherbacks are also the fastest-moving reptiles in the world, with recorded speeds in the water reaching 22 miles per hour—earning a listing in the 1992 edition of the *Guinness Book of Records*. We presume that during the day the animals dive deep in pursuit of their favorite food, the jellyfish. At night they ascend, again following their food source. As their diet consists almost exclusively of jellyfish, leatherbacks keep the jellyfish population in check.

Despite their unmistakable size, we know relatively little about these dinosaurs of the sea. The giants have populated the oceans of the world since prehistoric times, but we still don't know precisely how long they live. While some people think they live to be thirty, others believe fifty to be more likely and still others reckon on a hundred years. Leatherbacks are found in all tropical and subtropical waters, and thus have the widest range of all existing sea turtles. Their distribution area stretches in the north to Alaska and Norway; in the south they can be found off the southernmost tip of Africa at Cape Agulhas and the southern tip of New Zealand. One reason that the leatherback sea turtles are so successful in spreading throughout the oceans of the world is their body temperature. Although, like all reptiles, they are ectotherms (cold-blooded animals), they can maintain a body heat of up to 70°F above the surrounding water. The animals spend almost all their lives in the infinite blue of the oceans, with only the females returning every couple of years to the beach where they hatched or a sandy beach nearby to lay their eggs. Sometimes this means swimming thousands of miles from their feeding grounds back to their nesting beaches in tropical and subtropical regions. Worldwide, some sixty of these known nesting sites can be found on various beaches along the Atlantic, Pacific, and Indian Ocean coasts. The female leatherbacks come ashore at night to lay their eggs, crawling above the high tide mark to then, with great effort, dig holes in the fine sand using their flippers. On average they lay about 110 eggs, of which only about 15 percent don't make it to hatchlings.

After laying the eggs, the female leatherbacks carefully cover them with sand and return to the ocean. During the egg-laying season, the females return to the beach a number of times to dig further nests and lay more clutches. After about two months, the baby turtles hatch and, under cover of night, dig for the surface to begin a long and dangerous journey. As with other reptiles, the sex of the hatchlings is determined by the ambient temperature of the nest. A roughly equal mix of male and female hatchlings can be expected with ambient temperatures of 85°F. Higher temperatures increase the proportion of females, and cooler temperatures the proportion of males. After leaving the nest, the hatchlings head toward light, which on natural, undeveloped beaches leads them to the open horizon of the sea. Once in the water, the males will never leave it again, whereas the females will complete the cycle by later returning to the beach of their birth to lay their own eggs. The hatchlings, on their own once they have dug their way to the surface, are confronted by countless dangers. Already on the beach, hungry seagulls swoop down on them, and wild dogs, coyotes, lizards, and ghost crabs also get their fill. The hatchlings that make it to the water are still far away from safety, as there are other dangers lurking. In the sea they are on the menu of octopuses, sharks, and other larger fish. Very little is known about the young years of the sea turtles, so little that they are often referred to as the "lost years." We know that the animals follow migration routes for a number of years, and that these routes span whole oceans. Where exactly they remain or what they eat is still largely a mystery. A number of years later, and considerably larger in size, they emerge near their feeding grounds. For the past couple of decades, research groups have been gathering small pieces of the jigsaw puzzle from a variety of sources—shipboard observations, data of sea currents, tissue analysis, and satellite tracking.

More is known about the migratory routes of loggerhead sea turtles, especially the Florida population. The journey of the hatchlings starts on the beaches of Florida. After the hatchlings have struggled to find their way through the sand to the surface, running the gauntlet to the shallows begins. On a roughly 150-foot trail from the nest to the sea, the hatchlings scramble for their lives. The turtles, who have dodged hungry gull beaks and outmaneuvered the crabs, coyotes, and company, swim nonstop for a couple of days and nights to escape the danger zones of the shallows and reach the relative sanctuary of the warm Gulf Stream where they dive and drift. Once they arrive at the Sargasso Sea they find safety and rich food supplies of brown macroalgae (*Sargassum*) on the surface. *Sargassum* form thick raft-like clusters on which the hatchlings can rest and regain their strength. A wide diversity of organisms lives in the pelagic mats, including crustacean larva, barnacles, fish eggs, and other creatures that the hatchlings feed on. The juveniles allow themselves to drift with the current, thousands of miles toward Newfoundland and then toward the Azores. However, their route to the Azores is not direct, as they navigate around colder seawater areas to retain heat. They spend most of their time near the surface, using sunlight to warm up. From the Azores, they head back to the Caribbean, where they spend the next 20 to 30 years before the females swim back to their nesting beaches to lay eggs.

The loggerheads have a real superpower—they can orient themselves using the Earth's magnetic field. With the aid of their sense of magnetism, they can recognize the geographical latitude and longitude. Every region has its own typical geo-magnetic pattern, and the animals use this pattern to navigate the oceans and to find their way back to their birth beaches. Once they have found their way back, the females dig a nest in the sand and lay 23 to 180 eggs. After between 49 and 80 days, depending on the incubation times in the warm sands, the turtle babies hatch and head for the sea. The cycle is complete. The loggerheads' migration is one of the longest in the whole animal kingdom. They can travel almost 10,000 miles before reaching their birth beach on the coast of Florida—roughly the equivalent of flying from Vancouver, Canada, to Perth, Australia. Not all hatchlings reach adulthood, though. It is estimated that only one in a thousand turtles reach propagable ages. Gulls, sharks, and other predators are not their only problems; their greatest enemies are humans. Even though all species are protected by the Washington Convention—they are not allowed to be fished or killed, and trade in all sea turtle products has been prohibited since 1979—tens of thousands of the animals are killed by humans every year, whether by the development of hotel complexes at nesting sites, the stealing and trade of eggs, or the killing of adults for meat and the much-sought-after tortoiseshell of their shells. Additionally, every year countless animals die in fishing nets as bycatch, on longlines, and through pollution of the seas. Animal protection agencies throughout the world work on protecting the nests from poachers and escort the hatchlings to the sea in an attempt to assist the continued existence of the species.

It is not only sea turtles that are endangered. The largest marine animals, whales, have been and still are being hunted by humans. Hardly any other animal is so closely linked to myths and legends as these giants of the sea, and for hundreds

100

of years we humans have been linked, in one way or another, to these huge creatures.

Intelligent Singers

Whales are the largest existing mammals on this planet and, similarly to humans, their social structures are extremely complex. They have a large vocabulary and can even name one another. The various hunting strategies of the different species are clever and indicate a high degree of intelligence. Whales live exclusively in water, and there are some ninety species that can be split into two groups: the baleen whales (Mysticeti) and toothed whales (Odontoceti), a suborder that also includes dolphins. The first group filter plankton and small fish using comb-like baleen plates hanging from the upper jaw. This group includes most of the largest whale species, such as the humpback and blue whales. Toothed whales, not surprisingly, have teeth to catch their prey instead of keratin "beard" filters. All whales have a common tail fin—the fluke—which unlike other fish is horizontal and moves up and down when swimming. (Admittedly, classification of fish can be a bit confusing due to similarity in body shapes, but the fluke of a whale or the tail fin of a shark are clear distinguishing features.) The tail fin of sharks is vertical and moves back and forth while swimming. Other differing features that classify whales and dolphins as mammals and not fish is the fact that they are warm blooded, they are viviparous (they bear living young as opposed to eggs), they suckle their young, and they breathe air (which requires swimming to the surface, while fish absorb oxygen from the water via their gills).

Whales and dolphins are very communicative animals, which in their habitat is also very necessary. As mentioned in the section "The Songs of Fish," many aquatic animals rely on acoustic communication, as visibility under water is often limited and smells spread slower. Whales and dolphins use singing, whistling, humming, and a variety of other sounds to communicate. Their methods of creating sounds differ very much from our own. Humans communicate by forming sounds and words from the air flowing through our larynx by means of our vocal cords, lip movements, and tongue. Toothed whales, which include dolphins, narwhals, porpoises, and sperm whales, mostly communicate via high-frequency clicks and whistling sounds, and, more rarely, with a sustained sequence of tones. It is believed that the clicks are used mainly for echolocation (gathering acoustic information about the surroundings) and that they make use of different frequencies, as required, to determine how close or distant they are to their surroundings. The echoes of the clicks provide information about the distance to the object, which could be their quarry, as well as its shape, composition, size, and speed. The animals are also able to isolate their own echo soundings so there is no interference from the echoes of other individuals in their group. The whistling tones, on the other hand, are used for communication with each other. Even 4- to 6-month-old calves develop unique whistle sounds, which they use the most throughout their lives. Such sounds are the distinctive signatures of single individuals—they are used as unique identifiers. In a sense, the animals choose a "name" for themselves and, on the basis of their selected sound, they are recognized by others. In the Preface I described my encounter with about 300 spinner dolphins in the northern Maldives. Before I could even see the creatures, I heard their whistling and chirping sounds under water. This happened many times, both before and after the encounter I described. I didn't always get to see the animals, as sometimes they were too far away or visibility was poor. They, on the other hand, were very certainly aware of us. By using echolocation they could discern our numbers and progress. Probably, they were just chuckling among themselves about the slow, ungainly, much-too-loud shapes and their four long, skinny fins.

The sounds of the toothed whales originate in a special organ in their heads. The equivalent of the human nasal cavity, the phonic lips, are found between a number of air sacs. When air is squeezed through them, the sacs begin to vibrate. This vibration is channeled to the melon, where the clicking sounds are made and transmitted outside. The melon, a specialized structure made up of fat and connective tissues, is located on the rounded upper part of the toothed whale's head (their foreheads, as it were).

Baleen whales, on the other hand, have no phonic lips but a larynx that, unlike humans, have no vocal cords. In fact, we still do not completely understand how baleen whales formulate sounds. Even though it remains a bit of a mystery, the songs of baleen whales—some of them highly complex—are the focus of countless studies.

In addition to being the largest animal on Earth, the blue whale (*Balaenoptera musculus*) uses its deep voice to communicate with its fellow animals over the longest distances. Blue whales are often loners, but even they, every now and then, like to keep in touch with one another. Their sounds—low-frequency calls like grunts, drones, groans, and clicks—are mostly inaudible to human ears, but at 180 decibels are some of the loudest noises in the animal kingdom, traveling thousands of miles under water (just to give you an idea of the volume, a jet taking off registers around 140 decibels). Blue whales, together with the humpback whales, belong to the rorqual family (Balaenopteridae) and can reach impressive sizes. With lengths of between 75 and 110 feet, and weights of 100 to 180 tons, they are the largest known animals ever to have existed on Earth. The head is proportionally very long, taking up to a quarter of their entire body length, with two noticeable blowholes that, when seen from above, resemble the shape of a Gothic arch. Their blows (the exhaled air) shoot straight up, can reach heights of 40 feet, and look like thin spouts of moist, fine spray. Blue whales, like humpback whales, are filter feeders and swim, mouths wide open, through waters rich in plankton. Below their mouths longitudinal folds of skin run to the navel. These furrows can stretch fourfold, allowing whales to accommodate tons of water and krill. On the closing of the mouth, the folds contract and the water is forced out through the baleen plates, with the krill remaining and eventually swallowed. In this way, the whales hunt down up to 7 tons of krill a day.

The true opera-singing stars of baleen whales have to be the male humpbacks (*Megaptera novaeangliae*), whose songs consist of regularly recurring single verses. The verses vary from whale to whale and can be continuously repeated, sometimes for as long as half an hour. These songs are among the longest and most complex in all the animal kingdom, and consist of squeaking and whistling and deep harmonic tones. The humpbacks use their song to try to impress females or male rivals, repeating the tunes over and over again for a number of hours and sometimes even days. Well, actually, this hasn't been properly ratified. The various populations of humpbacks, found in all the oceans of the world, could have different dialects, just as we do in the various regions of our own countries. These minstrels can also produce different tones, maybe for defining their boundaries or for drumming together a hunting party. Speaking of hunting parties, the humpbacks have developed a really astonishing strategy involving a bubble net. A whale, or sometimes a group, swims in ever tighter circles beneath a school of fishes and while doing so blows a continuous stream of bubbles. This creates a "curtain" of bubbles—the bubble net—which prevents the fishes from escaping and forces them to concentrate in an ever-shrinking area. The whale then swims vertically up from below with mouth wide open, swallowing and filtering its prey in one big gulp.

I can recommend the Azores, a group of islands in the North Atlantic belonging to Portugal, as the best starting point for people in Europe who want to see humpbacks but don't want to travel too far. From March to October, you can see up to twenty different marine mammal species, from the diverse dolphin types to the largest of all whales: the blue whale. Another famous hot spot for whale watching is Vancouver Island. From March to October, humpbacks, orcas, as well as gray and minke whales can be seen regularly, as they come to feed in northern British Columbia. But here, too, please check that the tour operators really care about the welfare of animals. Humpbacks, in particular, are known for their curiosity, and it is not uncommon for them to swim up to whale-watching boats. If the boat's skipper isn't careful and doesn't keep distance, the animals can injure themselves on the boat's propeller. Humpbacks are especially easy to distinguish from their relatives, as their flukes have wavy trailing edges and the underside of the tail features black-and-white patterns. On diving, the fluke usually towers above the surface and is clearly visible. Experts use the fluke patterns to identify individuals. A blue whale's dive can last, as a rule, only 3 to 9 minutes, but has also been recorded at lengths of up to 45 minutes. When diving, the whale's back curves, giving it the appearance of a humpback. They have large knobbly heads and lower jaws, and noticeably long pectoral flippers, which are usually white on the underside and black on top. Humpbacks are real acrobats, and their 30-ton bodies can jump clean out of the water, returning with a mighty splash. This breaching is not performed to please tourists but likely serves as a form of communication between animals. Additionally, they often lobtail—slap the water surface with fluke or flippers—to draw attention to themselves or to frighten off rivals. But here too, as with the songs, we have more questions than answers. Regardless, it sure is spectacular. When snorkeling off the Bermuda coast, I was once lucky enough to swim near two adult humpbacks. Next to those giants I felt very fragile and insignificant. The pleasure didn't last long, as with three flips of their flukes they headed for the depths.

On a 10-week sailing tour with the Aquapower Expedition, which I cover in more detail in the "Endangered Blue Wonder" chapter, an encounter with a school of pilot whales lasted a bit longer. Many whales are very sociable, living in schools with a strict hierarchy. Unfortunately, I cannot say for sure whether the whales that we observed were short-finned pilot whales (*Globicephala macrorhynchus*) or long-finned pilot whales (*G. melas*), as they are very similar. After we left Bermuda and had already sailed a week through the North Atlantic toward the Azores, the weather became calmer. During that week we sometimes had to struggle with really rough seas, leaving us with hardly any opportunities to wash, let alone shower. And even though we were sailing on a relatively luxurious catamaran with freshwater showers, we didn't want to unnec- essarily waste our water reserves. I had the extra disadvantage of feeling sick when I was below deck for too long, so a quick sponging down had to do. Some noses will probably be wrinkling now, but miraculously none of us smelled bad. The salty sea air of the Atlantic seemed to have conserved, or rather, deodorized us. Once the weather stabilized, however, we took the opportunity to jump into the blue waters of the Atlantic, hundreds of miles away from the nearest mainland. The din that we created hopping wildly in and out of the water didn't go unnoticed for very long. After a little while, we saw gray dorsal fins breaking the surface, inducing us to quickly climb aboard our boat. Our worries were groundless, as it was a school of pilot whales, accompanied by a couple of dolphins, circling our boat out of curiosity. Roughly 20 of them remained near us for 15 minutes, eyes peeking out of the water to inspect us. Then we got too boring for them and off they swam. This encounter was really special—there was even a calf among them. Pilot whales belong to the toothed whale parvorder and are related to orcas or killer whales (*Orcinus orca*). Killer and pilot whales belong to the dolphin family (Delphinidae), which, admittedly, is a bit confusing. The "whale" in their names refers more to their size than their zoological affinity.

Killer whales are the largest species in the dolphin family and are easy to recognize by their high triangular dorsal fins and their high-contrast black-and-white body markings. The head, back, and fins are black; the underside as well as an elliptical spot behind each eye are white. The contours between black and white are well defined. The male dorsal fin can be over 6 feet long and looks like a sword cutting through the water. Males can reach lengths of up to 32 feet; females, and their dorsal fins, tend to be smaller. The killer whale, depending on habitat and nutrition, can be split into three main and differing groups (ecotypes) that mainly but not exclusively feed on: a) other marine mammals such as seals and other whales (transient); b) coastal fishes (resident); and c) pelagic fishes and octopuses (offshore). Killer whales can be found in all the oceans of the world and move around, usually in pods of several related species. The head of the group is always an old cow, as the group consists of her calves and "grandcalves." This base unit is then matrilineal and can link up with other closely related and not-so-closely related units to form a school. It is possible to know the familial relationships within a group by the different dialects. The more similar the utterances, the closer the relationship. These sounds, just like the sometimes extremely complex hunting techniques, are passed on from the mother to her calves. Killer whales are alpha predators and, apart from humans, have no natural enemies. The various pods employ a variety of hunting tactics, depending on the habitat and the food supplies. Killer whales in the Antarctic, for instance, swim together in a combined effort toward an ice floe where one, or more, of their prey is resting. The act of swimming in unison creates a wave that, at best, washes over the ice floe, flushing prey like seals or penguins off the floe and into their open mouths. Killer whales that specialize in hunting other mammals don't communicate by echolocation even though they are well capable of doing so. It is believed that they don't want to unnecessarily advertise their presence. Pretty smart! In addition to smaller animals, killer whales will sometimes go after great whites and tiger sharks and even whales. In 2012, a number of great whites were washed up on the beaches of Gansbaai in South Africa, all missing their livers. The livers hadn't been eaten as carrion by other animals after their deaths but rather they had been killed specifically for their livers, which had been extracted with almost surgical skill. The rest of their bodies were more or less intact. It was thought that killer whales were keen on the livers as they are rich in fish oils. News of the attack had obviously spread, as the great whites disappeared from the area for a while. Diving excursion operators in Gansbaai, which labels itself the great white shark capital of the world, reported zero sightings, and divers traveling there specifically to see the great whites had to return home disappointed. Sperm whales are also prey for killer whales. When attacked, sperm whales form a circle around their calves and the weaker members, with their heads pointing toward the middle and flukes on the outside, and try to drive off their attackers with their powerful flukes. Even though killer whales take on sperm whales, the latter are also notorious predators.

Sperm whales (*Physeter macrocephalus*) are the largest toothed whales on Earth, and thus pretty terrifying. Their name is derived from spermaceti, a waxy substance found in their head cavities, which was used in oil lamps and made sperm whales a prime target for the early whaling industry. Sperm whales inhabit all five oceans, and you can find them even in the Maldives. Not that I have ever seen them there, but we had an exhibit of a sperm whale skeleton in our Eco Centre on Kuramathi. The animal was found dead on the beach of another island a number of years ago and then was buried in the sand. After it had completely decomposed, the remaining skeleton was excavated, cleaned, and professionally prepared as an exhibit, and can be admired at the Eco Centre to this day. The natural life span of sperm whales is around 60 years. Sperm whales reach an average length of a little more than 50 feet, with the females being somewhat smaller and lighter than the males. Their extraordinarily large heads are worth mentioning, as they constitute roughly one-third of their total body mass.

Odontoceti differ from baleen whales, not only by having teeth but in the number of blowholes in the middle of the head. Unlike their filtering relatives, they have only one blowhole instead of two. All species spout through the blowhole after surfacing from a dive—the sprayed fountain of the blow is easily visible on a good day. The direction and the shape of the fountain-like spray vary from species to species and help whale watchers with identification. As mentioned, sperm whales can dive extremely deep when seeking food. In depths of 1,000 to 2,600 feet, in the mesopelagic zone, they hunt their favorite food: octopuses. Examinations of the stomach contents of dead sperm whales have found remains of deep-sea species, indicating that they can dive considerably deeper, even to the dark depths of 10,000 feet. Such a dive can last over an hour, but then the hunter has to return to the surface and retank with some fresh air. Sperm whales have an unmistakable body shape—especially the males, with a noticeably large, rectangular head housing a brain weighing up to 20 pounds, the largest in the animal kingdom. In comparison, a human brain weighs 3.3 pounds on average, depending on sex and height. The head takes up almost a quarter of the total body length in size, and the body is covered in dark, wrinkled skin, akin to the surface of a prune. In addition to the enormous brain, the oversized head houses the soft wax of the spermaceti oil in a cavity the size of a car. Its precise function is not yet known, but it is thought to be linked to echolocation and buoyancy. Another possibility is that it gives the bull whales more stability and power when they are ramming each other. Ramming isn't restricted to rival males, however. There are also documented cases of sperm whales deliberately ramming ships (at least one whaleship capsized and sank after being rammed by a sperm whale).

Sperm whales, which in the past had been hunted almost to extinction, are today suppliers of a substance for which large amounts of money are paid—sperm whale feces. Yes, you read it correctly. Sperm whale poo is worth its weight in gold and is traded (under the table) worldwide as one of the most precious substances. It is quite possible that you have come across it, perhaps even placed a little dab on your wrist or a quick squirt in your cleavage. We are talking about ambergris, a dull-gray waxy substance from the rectum of sperm whales with digestive problems. As you may know, sperm whales like octopuses very much. The indigestible bits of octopuses, such as the beaks, present risk of injury to the sperm whale's digestive tract. Normally, sperm whales vomit regularly to eject these sharp-edged food remains. Bits of the beak, however, often remain in the gastrointestinal tract, enclosed by an oily clump to protect the sensitive intestinal walls. This clump is ambergris, and we still do not know exactly how it is produced deep inside the guts of sperm whales. Humans come into contact with this precious substance either through natural processes when it exits the whale's anus or when the whale dies and the ambergris pops up to the surface after decomposition. In both cases, it is possible that the ambergris is washed up on the beach as flotsam, where knowledgeable beachcombers scoop it up and sell it. During its journey across the sea, ambergris changes in constituency, color, and smell through exposure to salt water, light, and air, transforming over the course of time to one of the most expensive ingredients for perfumes in the world. Because this substance is extremely rare, it is also produced synthetically. Ambergris is only used in particularly expensive perfumes, under a cloak of secrecy, as trade in sperm whale products is forbidden according to the Washington Convention and European Council regulations.

In addition to whales and dolphins, another hunter swims the pelagic zones. Believe it or not, fish can also be successful and much-feared predators, and the fish that impress humans the most is, without doubt, the shark.

Hunted Hunters

Sharks are fantastic and magnificent creatures whose reputation, unfortunately, has been greatly damaged by films like *Jaws*. Despite this, no one can really take away the fascination that these elegant hunters exert. It's interesting to note that the two largest shark species—the whale shark and the basking shark—don't feed on people swimming in the water but on the tiniest creatures in the sea: zooplankton. Other shark species exclude humans from their menu, too. The Florida Museum of Natural History houses a databank, the International Shark Attack File, that tracks all shark attacks, and they update it regularly. In 2018, there were 130 interactions between humans and sharks worldwide, and only 5 were lethal. In fact, more people die from mosquito or dog bites than from attacks by these villains of the sea. The World Health Organization reckons

diseases like malaria and yellow or dengue fever are responsible for millions of fatalities a year. And swimming with your
pet dog (mine, too) is potentially more hazardous than being
in the water with a shark—dog saliva contains dangerous bacteria that, if transmitted to an open wound, could in worst
cases lead to amputation of the infected limb or even death.
Of course, I'm not trying to defame pet dogs—mine would
be pretty unhappy about that—just simply pointing out that
many more fatalities are caused by household pets, accidents
at home, mosquitoes, cars, etc., than by sharks. When I was
in the Maldives I came into contact with sharks in the water
almost daily, and I still have all my body parts. And despite
being in the water regularly when I had my period or bleeding
grazes on my knees, these "bloodthirsty" creatures were totally
disinterested in me being their next snack. They weren't even
interested in trying me! The same applies to my diving friends,
acquaintances, and colleagues. During bad weather I even felt
safer under water than on land: When walking to the office or
riding my bike, I would give swaying palm trees a wide berth
to avoid the whole bunches of coconuts that would frequently
come crashing down to the ground. When it was properly
windy, the nuts would fly around like grenades. I know more
people who've experienced broken noses or various bumps
and lumps from falling coconuts than have even been glanced
at by a shark. It sounds cliché, but it really is true. On the other
hand, humans kill sharks by the millions every year, many of
the sharks left in agony as their fins are cut off while still alive.
After finning, as it is called, the sharks are thrown back into the
sea where they suffer a wretched death. And for what? To end
up in tasteless soups or to allegedly increase male potency?
A research team led by shark biologist Samuel H. Gruber,

founder of the Bimini Biological Field Station in Miami, calculated that more than 100 million sharks are killed by humans annually, and it believes that this estimate, based on figures from the fishing industry, is conservative. That's 11,417 sharks per hour. According to the authors of a study published in the scientific journal *Marine Policy* in 2013, this figure is more likely 273 million a year. Try to let those numbers sink in, if you can—even then, they are simply incomprehensible.

Sharks are not bloodthirsty creatures but rather of vital importance to the marine ecosystem. They stand, after killer whales, at the top of the marine food chain and ensure that a natural balance exists between fish populations and their surroundings. Imagine for a moment a tropical coral reef without sharks. Within a very short time other predators that had previously been kept at bay by sharks would decimate the population of herbivorous fishes. This would mean that algae, which had once been grazed by the plant eaters, could now expand unhampered and overrun the coral, which eventually would die. This is not in the realm of theory. It has been scientifically proven that a lack of sharks is damaging to the health of the sea.

North Carolina has gained a sad notoriety. Overfishing of large shark species such as hammerheads, bull sharks, and great whites, which feed mainly on other cartilaginous fishes like rays, led to an uncontrolled explosion in the ray population. Between 1986 and 2000, the population of hammerheads drastically shrank by 89 percent due to heavy overfishing. Hammerhead sharks specialize in hunting rays, incapacitating them by striking them to the seabed with their hammer-like heads before finally eating them. After the threat from the large predators was removed, ray numbers erupted, in particular those of the cownose ray (*Rhinoptera bonasus*), which swelled to a population of over 40 million. By 2004, this army of rays actually managed to completely ruin the local 400-year-old mussel-fishing industry. The fish simply gobbled up the reserves. When humans exterminate whole segments of the ecosystem, particularly alpha predators such as sharks, the balance between the species tilts, causing a domino effect throughout the whole food chain. Some of these effects can be damaging to humans in the long run.

Something has to be done pretty quickly about the bad image of sharks to prevent extinction. Luckily, an increasing number of responsible tour operators are interested in animal welfare and aim to introduce humans to sharks in their natural environment. SharkSchool, for instance, offers workshops run by the respected shark researcher Erich Ritter. Participants dive and snorkel among sharks off the island of Eleuthera in the Bahamas in order to learn about the behavior and biology of the creatures. This is a practical approach to protecting sharks—once people learn to understand sharks, they lose their fear of them and their desire to protect these fascinating animals grows.

Thinking about the great white shark (*Carcharodon carcharias*) still sends shivers down most people's spine and provokes the theme song of Steven Spielberg's 1975 movie *Jaws*. According to Spielberg, no boat, no swimmer, is safe from these teeth-brandishing monsters, as great whites will destroy boats just to get at some tasty humans. Inevitably, if you swim unwittingly in the sea, a shadow will shoot up from the depths, sink its teeth into you, and drag you down in a swirl of blood and bubbles. I, too, saw the film as a teenager and was scared stiff. From then on, I was terrified of sharks—but only until I dived with my first large shark, a great hammerhead (*Sphyrna mokarran*). Of course, I can't claim that I was wasn't at all anxious, but my anxiety quickly turned to absolute fascination in these creatures and that has remained to this day. Worldwide, there are 500 known shark species, and they can be found in all marine environments—in coastal waters and the open seas, in the Caribbean and in the Arctic, even in fresh and brackish waters. And yes, there really are sharks in rivers and lakes. But you don't need to be scared. You will not bump into a shark in Lake Constance or Yellowstone Lake or even Cedar Creek Lake or Swan Lake. For that, you need to visit the rivers of India, Southeast Asia, and parts of Australia, where rivers flow into the sea. But even in these places the likelihood of encountering the Loch Ness Monster is greater than coming across one of these shy creatures. There are currently only five known freshwater sharks, but because they are so seldom sighted little research has been done on them.

Glyphis is a genus of river shark in the Carcharhinidae (requiem shark) family. It is believed that they mainly feed on fish and that the females are live-bearing. River sharks can probably reach lengths of up to 10 feet, but since to date so few adults have been sighted it can't be properly substantiated. Whether we will ever learn more about the secret life of these animals is uncertain, as the human impact of overfishing, pollution, and the shrinking of their habitats have made them endangered species. Luckily, we do know a little bit about them. The speartooth shark (*Glyphis glyphis*), for instance, is found primarily in river areas in the Indo-Pacific region and preferably in the mangrove-lined rivers of northern Australia. From an Australian population of speartooths, it is known that the juvenile sharks, after birth in the estuaries, migrate up to

50 miles upstream from October to December. At this distance from the sea, the salinity is much reduced, and here they spend their first 3 to 6 years. Since the annual monsoons supply the rivers with much fresh water, the juvenile sharks can migrate downstream to areas were the salinity reaches the levels that they prefer. It is believed that adult spearheads migrate to the river mouths to breed and then return to the coastal water regions of northern Australia.

Another river shark species is the extremely rare Ganges shark (*Glyphis gangeticus*), which lives in large rivers like the Ganges in northern and eastern India. Very little is known about these sharks, and they are often mistaken for the more common bull sharks (*Carcharhinus leucas*), which can also be found in the Ganges. Bull sharks inhabit warm shallow waters in many parts of the world along the warm coasts of Africa, Asia, and North, Central, and South America. As their preferred habitat is the murky shallows, they, together with the great white and tiger sharks, are responsible for the most human interactions. Bull sharks are aggressive defenders of their territory, with practically zero tolerance of intruders. It is thought that many attacks attributed to the great white were, in fact, instigated by bull sharks. Bull sharks are not found exclusively in salt water. Due to a special physiological feature, they can adapt to wide ranges of salinity, from the brackish waters at the mouths of rivers to fresh water upstream. The metabolism in their kidneys can adapt to their respective surroundings, enabling them, at least theoretically, to spend their whole lives in a freshwater environment. Lake Nicaragua, the largest inland lake in Central America, near the border to Costa Rica, actually has a population of bull sharks that were erroneously identified as their own subspecies, *Carcharhinus nicaraguensis*. However, these sharks don't spend all of their lives in the lake; they migrate down the Rio San Juan in Costa Rica to the Caribbean Sea. Data from tagged bulls recorded that they had taken 7 to 11 days to travel the 125-mile stretch of river from the lake to the sea. Sadly, the population of Lake Nicaragua bulls is threatened by overfishing, and even the protective measures taken by the Costa Rican government don't seem to be enough to rectify the situation. Bull sharks have been recorded in many tropical rivers and lakes in Australia, New Guinea, the Philippines, Asia, and Africa, and all parts of the Americas. Bull sharks have also been sighted deep inland, 2,300 miles up the Mississippi and 2,600 miles up the Amazon in the foothills of the Peruvian Andes. All in all, bull sharks are large and stout, with wide snouts. The females are, on average, 8 feet long and weigh 290 pounds, although a 13-foot specimen has been caught, and the heaviest-recorded creature was all of 700 pounds on the scale. Bull sharks are, on average, 7.4 feet long and weigh 210 pounds. Like most species of sharks, the bull sharks have a number of rows of teeth so that lost ones can always be replaced. Some people believe that some of these teeth—or rather, their owner—was responsible for the 1916 shark attacks off the coast of New Jersey, where within 11 days 4 people were killed and a further person injured. Others think that a female great white was the culprit. This series of attacks was the inspiration for Peter Benchley's 1974 novel, *Jaws*. He later regretted the impact of his novel, and from then on involved himself in shark and marine preservation. Just one year after the novel was published, Steven Spielberg's film adaptation was released, which triggered the widespread fear of great white sharks and their relatives that continues to this day.

The great whites (*Carcharodon carcharias*) are, by nature, inquisitive creatures and have been observed numerous times swimming alongside boats while sticking their heads out of the water to see what was happening. Spielberg would likely have translated this as calculated preparation for an imminent attack on the helpless people on board. This curiosity actually does pose risks for both man and beast. Roughly a third to a half of just over 100 shark attacks recorded worldwide each year can be traced to great whites, but most of the attacks are not fatal. It is thought that many of the attacks happen out of curiosity—the shark just takes a "sample" bite to confirm, quite rightly, that we are not supposed to be on the menu. Such attacks, admittedly, can be traumatic, and sadly in some cases they do end fatally, after all a shark's mouth is not the same as, say, a cat's mouth. Still, sharks do not attack with the intent of eating us; that is just the artistic freedom of a film director. The hysteria whipped up by the film *Jaws* has had a devastating effect on the great whites, and their numbers have dwindled since the 1970s to such an extent that they are now categorized as "threatened." Estimates reckon the global population stands at about 3,500. The chances of accidentally encountering one are accordingly very slim. If, however, you do come across one, its size alone is enough to command respect and stoke fears. After the whale and basking sharks, the great white, with its 13- to 20-foot body length and weight of around 4,500 pounds, is the third-largest shark species. From the bite marks found on whale carcasses and occasional catches, it is believed that they can reach maximum lengths of up to 23 feet plus, making them one of the world's largest predators. The adult females, as with many of the shark species, are larger than the males. Their slate-gray bodies perfectly match the rocky sea beds of coastal waters. (Their name is actually derived from their white undersides.) Their streamlined torpedo-like bodies are perfectly suited to swimming, and they can reach top speeds of 35 miles per hour. Great whites are cosmopolitan, and they can be found almost all over the world but prefer coastal waters as well as open seas with temperatures between 54° and 75°F. Larger concentrations can be found off the coasts of South Africa, northeastern United States, California, Japan, Chile, Oceania, and *maybe* the Mediterranean Sea. But expect to be disappointed if you pack your diving gear hoping to see one there. The appearance of great whites in the Med is a little bit mysterious. It is theorized that the Mediterranean serves as a sort of kindergarten, as only juvenile animals of 3 to 4 days old have ever been caught there. No one really understands where the adults are and what they find to eat in the overfished waters of the Med. Generally, great whites are opportunistic hunters, but they don't turn up their snouts to dead animals. Their prey ranges from invertebrates, to tunas, to smaller marine mammals and other sharks. Measurements have demonstrated that the great whites have the greatest bite force of all animals, with larger specimens registering biting forces of some 4,095 pounds per square inch.

Not all sharks feed exclusively on other animals. Apparently, the word is out about the health effects of vegetables. Although we have yet to discover vegan varieties, recently it has been discovered that there is a shark that values its greens. The bonnethead or shovelhead shark (*Sphyrna tiburo*), a roughly 3-foot-small member of the hammerhead genus living in tropical waters off the U.S. coast, has added seagrass to its menu—not as an exception but actually large amounts of it. By feeding the bonnetheads seagrass and octopuses for 3 weeks, research teams at the University of California succeeded in demonstrating in laboratory experiments that these sharks are the only known omnivores of their kind. In the experiment, 90 percent of the diet consisted of seagrass and only 10 percent octopus. Despite the almost completely vegetarian diet, the bonnetheads gained weight and were otherwise perfectly content. The teams also found out that the sharks digest and metabolize seagrass as well as any green sea turtle (*Chelonia mydas*), which graze whole beds of seagrass. But the diet is not the only unusual thing about this species. Their propagation is also unique. One female specimen kept at the aquarium at the Henry Doorly Zoo in Omaha gave birth to a young female without any assistance at all from a male. According to the conclusions of the research team that conduced various tests, when there is a lack of males, the bonnetheads can reproduce asexually. And there they are not alone; this form of reproduction has been observed with another species of shark. Leonie, a female zebra shark (*Stegostoma fasciatum*) living without a male partner in an aquarium in Townsville, Queensland, Australia, also gave birth to offspring without the help of a male.

The fastest sharks are the shortfin mako sharks (*Isurus oxyrinchus*)—the marine equivalent of cheetahs, the fastest land animal. The shortfins can reach top speeds of 47 miles per hour and, aside from being fast, have plenty of stamina. One tagged animal swam 1,250 miles in 37 days, with average daily distances of 34 miles. Their speed means that they are able to hunt other fast fishes such as tuna, mackerels, swordfishes, and even other sharks. In exceptional cases, these slender sharks can grow to 13 feet long and weigh 1,100 pounds, but generally are around 10 feet long and between 132 and 300 pounds. Shortfin makos feel at home in tropical and subtropical waters, with a preferred temperature range of 62° to 73°F and depths of up to 500 feet. Like their cousins the great whites, they belong to the mackerel shark family (Lamnidae). With their large, black button eyes, they could look really cute if it weren't for the pointed snout and the even more pointed thin teeth, which can be seen even when the mouths are closed. Unfortunately, because of their speed and strength, anglers consider them "worthy opponents" whose stuffed heads or teeth are thought to prettify living rooms. But sport fishermen are not the only ones who pose a threat to these agile hunters; the deep-sea fishing industry is also a threat. In the meantime, their conservation status is endangered.

Amazing Appendages— Swords, Spears, and Wings

Mako sharks are not the only fish popular with sport fishermen. Their prey, such as the marlins, are also much sought-after. Many people visit the Maldives specifically to catch big fish. Big game fishing is very popular there, particularly among male adventurers. Above all, it is the billfish—which include swordfish, sailfish, and marlins—that are popular trophies. All three are large, fast, and strong fishes who don't shun a fight with anglers but who, sadly, lose all too often. I still cannot understand the appeal of killing animals for trophies and not for food. These animals are more beautiful in water and that is where they belong, not tacked to a living room wall. For visitors, the most coveted prize was landing an Indo-Pacific sailfish (Istiophorus platypterus) complete with souvenir photos shot on board or at the quayside. Sailfish are related to swordfishes and marlins, which also have long dart-like pointed bills, and are very fast swimmers. They live in pelagic zones in depths of up to 650 feet but prefer the upper reaches, with temperatures between 70° and 82°F. Sailfish are truly impressive in a number of ways. Their outward appearance alone is striking—their dorsal fin, which at its highest point it is wider than the body, looks like a sail, or the spiked-up hair-style of a punk, rising up from their backs. It is speculated that this sail-like fin serves as a cooling/heating system because it contains numerous blood vessels and, when breaching the water surface (before or after fast-swimming sequences), it dis-plays "sail-raising" characteristics. And the punks of the sea can swim, even though there are disputes about top speeds—some say over 60 miles per hour; others 28 miles per hour. Sailfish are the Speedy Gonzaleses of the sea. They zoom through the oceans like torpedoes, and researchers have speculated long and hard about the use of their bills. Since a dart or spear is mainly used to impale or jab opponents or prey, it was origi-nally thought that sailfish must use their bills to impale their prey. This theory was revised once the stomach contents of sailfish were examined and no visible signs of external injury were found among the many fish inside. Apparently, sailfish use their bills more like a blender than a weapon. The roughly 8- to 13-foot creatures stalk their quarry, for instance a school of sardines, from behind, slashing their bills sideways, to and fro, extremely quickly to agitate the water. The movements, which can reach speeds of 12 miles per hour, are not visible to the naked eye and can only be recorded using high-speed cam-eras. Fish that get caught in this motion become disoriented and are snapped up by the sailfish. Interestingly, when large numbers of sailfish are hunting in a pack, only one member of the group slashes at the herded school, possibly to avoid injuring each other.

Fish are obviously water dwellers and swim to get from A to B. This is true and proper for almost all fish, but some finned creatures have achieved the impossible and can fly. Flying fish (Exocoetidae) are present in all oceans but particularly in tropical and subtropical regions. They live in the epipelagic zone, an extremely rich habitat teeming with life. The epipelagic zone, however, also conceals many dangers for smaller fishes, as they are hunted by pretty much anything bigger than they are. All 45 varieties of flyers are relatively small, none larger than 18 inches. Flying fish have long, slim, streamlined bodies to minimize wind resistance. Their eyes are quite large in relation to body size, giving them good sight both above and below the waves. Under water they don't dare swim deep, as this negates their greatest skill. They spend most of their time close to the surface with their wings firmly pressed against their bodies. When these small fish are hunted, they catapult themselves out of the water, gliding with outstretched fins above the surface. Of course, in order to fly these animals need special adaptations—compared to other fish the considerably larger pectoral fins sit much higher on the body and are supported by well-developed chest muscles and a kind of strong shoulder strap. The animals, depending on species, have 2 or 4 wing-like fins that help them shoot out of the water and glide above it. Unlike birds, their "wings" do not flap but remain fixed once in position. If the creatures feel threatened, the tail fin begins to flap under water to gain extra thrust, beating more than 50 times per second to reach the right speed to propel themselves out of the water. Once airborne, they can glide parallel to the surface for about 45 seconds, covering distances of up to

1,300 feet at an average altitude of 5 feet. When they find they are losing height, they sometimes slap the water surface with their tail fins to extend the flight. When they lose momentum, they fold in their fins close to the body and dive back into the ocean. Occasionally, they reach top speeds of 45 miles per hour and heights of 16 feet. We were able to view the flying skills of these extraordinary creatures almost every day during our sailing expedition, as they are found in tropical and subtropical regions of the Atlantic, Indo-Pacific, and in the Mediterranean. In the mornings, we often found a few stranded flying fish up on the deck of our catamaran—luckily, we were able to return most of them to the sea alive.

Now that we've taken a peek across the wide expanse of the oceans, we'll dive into the area that still holds secrets—the largely unresearched deep sea.

THE SECRETS OF
THE DEEP

~~~~~~~~

THE DEEP SEA is a place full of secrets. Full of great big secrets! In the dark recesses of the oceans, the organisms seem to have come from another dimension or escaped from Jurassic Park. Only in recent decades have we succeeded, with a lot of effort and bit by bit, in shedding light on some of its mysteries. Although the deep sea makes up the largest portion of the seas and oceans by far, we currently know less about its depths than we do about the moon.

The reason is that, similar to the terrestrial world, this hostile world is covered in mountains and trenches. The deepest point on Earth, the Mariana Trench, is just over 36,000 feet, and is thus deeper than Mount Everest is high (the world's highest mountain is "only" 29,029 feet). The deeper you go, the higher water pressure climbs. At sea level, each square inch of our body surface is subjected to a force of 14.6 pounds. The pressure increases by 1 pound-force per square inch for every 33 feet of water depth. At 1,640-foot depths, for example, the pressure will be 500 times greater than the pressure at sea level.

While the pressure increases, the temperatures drop, fluctuating between 30° and 39°F—all in complete darkness. It is hardly imaginable that, under such extreme conditions, any organism could call this place home. Life at these depths feed mostly on what falls from above: detritus, also called marine snow, sinks in flakes from the sunlit zones down to the depths. These snowflakes consist of decomposing animal or vegetable materials. Everything is valuable nutrition to organisms of the deep, from dead whale cadavers to the excrement of zooplankton and algal remains.

Other places also support deep-sea life—underwater thermal vents are proper oases in this otherwise cold, low-oxygen, dark world.

Life at the Extremes

Volcano-like forms, similar to smokestacks, protrude through the darkness spouting warm water. These hydrothermal ocean vents are mostly found on mid-ocean ridges and usually at depths of between 3,250 and 13,000 feet, where they appear in clusters. Mid-ocean ridges are volcanically active mountain chains in the deep sea that run through all the oceanic basins and stretch adjacent to continental or lithospheric plates. In these ridges, minerals in the hot water are released and precipitate in the 35°F waters. "Precipitate" means that the minerals, principally sulfides but also other salts of iron, zinc, copper, and magnesium, flocculate as fine particles. These fine particles compound on the ground in the course of time and form the characteristic smokestack vents, which reach an average height of 65 to 80 feet. The clouds of particles resemble smoke coming out of a chimney, which is why they are sometimes referred to as "smokers." If the cloud is rich in iron sulfide (e.g., pyrites), then the smoke is black and they are then, understandably, called "black smokers." The water of a black smoker reaches temperatures of 570° to 750°F because the cold circulating seawater is drawn in and pressed through faults and crevices several miles below the ocean bed, flowing past chambers full of boiling magma and heating up. The pressure built up in the process allows the water to shoot out of the chimney vents like a fountain.

Scientists first discovered hydrothermal oceanic vents 200 miles northeast of the Galápagos during the Galápagos Hydrothermal Expedition. On February 17, 1977, a deep submergence vehicle, Alvin, manned by three scientists and equipped with measuring devices, cameras, and gripper arms, was launched by the mothership and dived to a depth of 8,250 feet. Down below, on the abyssal floor, the crew were in for a surprise. Alvin's temperature sensors registered a difference in water temperatures indicating hot springs. In the icy deep seawaters, they were getting readings of 46°F, and then they entered a different world, a world of white clams clustered in an area roughly 165 feet wide. The warm water coming out of cracks in the lava landscape transformed the surroundings into a blue, milky, shimmering cloud floating above the footlong clams. They also discovered brown clams, ghostly crabs, and a purple octopus that likely preyed on other animals near the vent. Incredibly, at a depth of some 8,250 feet, they had discovered the first hydrothermal vent, and the area was teeming with life. They named the area where they found the vent "Clambake 1." A couple of dives and further vents later the scientists discovered a hydrothermal vent with such a diversity of life that they named it the "Garden of Eden." The fascinated scientists observed numerous different organisms bustling on and around the black smoker, and all in pleasant temperature of 63°F. In this underwater paradise they saw, for the first time, white-stalked tubes with red tops looking like a field of flowers swaying in the wind. These were later identified as *Riftia pachyptila*, giant tube worms. Tightly clustered, these bizarre creatures were hugging the vent site. Their white stalks grow as thick as an arm and up to 10 feet long, and feature a bright red plume that, if threatened, retracts into the stalk. The tube worms shared their habitat with small white crabs spooking around like ghosts, snails sitting on the lava, and tiny fishes. The scientists were overjoyed and felt like Columbus discovering their very own new world. They took samples of animals and water to the surface to examine in the lab on the mothership in an effort to determine how the organisms managed to survive in the inhospitable surroundings. They stored the specimens in some strong Russian vodka because they had run out of formaldehyde. When analyzing the water, they were very nearly overcome by the smell of bad eggs. The smell of hydrogen sulfide confirmed their theory that the warm water was seeping out of fissures near magma chambers. At these sites, the heat of the seawater converts the abundant supplies of sulfates to hydrogen sulfide. The hydrogen sulfide, together with the hot water streaming out of the fissures in the lava, mix with seawater rich in oxygen to create the blue clouds above the vents. Bacteria and other microorganisms metabolize the hydrogen sulfide emitted by the hydrothermal liquids, obtaining energy or their nutrition, then grow and flourish. The process is called chemosynthesis and is akin to photosynthesis under water. As was later discovered to the astonishment of the scientists, this form of energy extraction is useful to other organisms. *Riftia pachyptila* have no digestive system, no mouths, no bowel outlets, but rather very interesting organs called trophosomes—collections of enlarged cells containing sulfur bacteria covered in minute blood vessels. Evidently, *Riftia* live on nutrients produced by the bacteria exclusively and, in return, the tube worm provides the necessary link to hydrogen sulfide and oxygen. The resident clams, too, live in symbiosis with hydrogen sulfide bacteria. The bacteria receive filtered sulfide water from the clams and in exchange provide the clams with nutrition. The scientists had not only discovered the black smokers but also how underwater oases like these could support so much life in such an inhospitable environment—a really pioneering discovery that is in no way inferior to Columbus discovering America.

As well as black smokers, there are other kinds of hydrothermal vents: white smokers that emit clouds of light-colored sulfates such as gypsum and anhydrite. Waters emitted by these vents is only 100° to 200°F, so considerably cooler than their black relatives. These kinds of deep-sea hydrothermal vents were first discovered by research teams from the United States and Switzerland in the mid-Atlantic at a depth of 2,600 feet in 2000. The team found a collection of white vents up to 200 feet high and named this bizarre underwater formation the "Lost City." Two years and one expedition later, the first secrets of this ghostly underwater city were unveiled. The vents of the white smokers are created by a chemical reaction between the iron-rich minerals of the rocks of the Earth's crust and the surrounding waters. Put simply, serpentine is formed from the release of water in the reaction that then, bit by bit, piles up to form a kind of white chimney. In the process, caustic alkaline solutions with pH values between 9 and 11 are created. The amazing thing is that the "Lost City" has existed for roughly 100,000 years, and it is still continuing to grow. Actually, it is thought that the process of serpentiniza-tion is as old as the oceans of the world. As is right and proper in a city, the "Lost City" is densely populated but by bacteria. These bacteria, called chemoautotrophic bacteria, convert sub-stances seeping from the Earth's interior into energy. Deep-sea hydrothermal vents like the ones in the "Lost City," with their primeval bacterial communities, are considered a possible source of life on the planet: the Earth supplies chemical energy that the vents then convert to organic life to create biomass. As primary producers, these thermophile chemoautotrophic bac-teria are right there at the beginning of a long food chain.

Although most technology today has a short life, this appar-ently doesn't apply to *Alvin*. The submersible, with room for three people, is still operating. As recently as 2014, a research team searching for hydrothermal vents made another sensa-tional discovery. During a dive 150 miles off the Pacific coast of Costa Rica, at a depth of around 10,000 feet, the crew stum-bled across about 100 immobile pale blue forms. On closer inspection, they recognized octopuses belonging to the genus *Muusoctopus* dotted around the sea bed. The octopuses appeared to be inverted, or turned inside out, a common pose among females when brooding. The researchers had discovered an octopus nursery: the animals were all females protecting their eggs. Because the bed of the deep sea mostly consists of sludge and marine snow, animals can't lay eggs on its soft surface, so good breeding grounds are rare. In this location, called the Dorado Outcrop, however, deep-sea vents were bubbling up through cracks in the hard volcanic rock. This hard foundation evidently provided breeding octopuses with the best possible conditions, at least at first glance. The researchers noticed that the animals were showing signs of stress, as the females had wrapped their arms around their bodies. Since octopuses usually live in colder water, which is richer in oxygen, the scientists thought the animals might be protecting their eggs from the much-too-warm water. The team reported that the eggs didn't show any signs of developing embryos and supposed that the octopus moms didn't have the luxury of being able to choose a better breeding spot. Early in 2018, the crew of the American research ship E/V *Nautilus* discovered the largest gathering of octopuses—over 1,000 *Muusoctopus robustus*—off the Californian coast. The breeding octopuses had settled everywhere on Davidson Seamount, and here, too, the bodies were inverted. The magmatic waters seeping out of fissures appeared to be warm: the videos taken seem to show heat shimmers around the octopuses. Even though no temperature measurements were validated, the researchers believe that the octopuses chose this site as a brooding ground because of the water temperature. This, however, contradicts the investigations in 2014 off Costa Rica, and the American team now believes that another factor (not the heat) was behind the damaged Dorado eggs. How exactly the animals select their brooding grounds, whether warm water or an unknown factor, remains, for the time being, a secret of the deep.

In addition to providing a nursery for octopus kids, deep-sea vents also provide for the offspring of *Bathyraja spinosissima*, the Pacific white skate, of the family Arhynchobatidae or softnose skates. In June 2015, a research team led by Pelayo Salinas-de-León and assisted by *Hercules*, a remotely operated underwater vehicle (ROV), investigated some black smokers in the Galápagos Marine Reserve. At depths of between 5,250 and 5,500 feet they found a total of 157 skate egg cases lying embedded on lava rock, no more than 500 feet away from two active hydrothermal deep-sea vents. Using the robotic arm, the team gathered four of these 4-inch-long egg cases for further examination in the lab. The softnose skates had apparently been using the heat of the thermal vents to incubate the eggs. The research group believe that this behavior is aimed at shortening the incubation time of the deep-sea skates' embryos. A skate of the same genus, B. *parmifera*, found in the Bering Sea, has an incubation time of an impressive 1,290 days at water temperatures of 40°F. Assuming the surrounding water temperature for B. *spinosissima* was 37°F, incubation time would be around 1,500 days. They concluded that warmer temperatures would considerably speed up the development process.

The female deep-sea octopuses of the *Graneledone boreopacifica* species can only dream of speedier incubation times. The small, cute, deep-sea octopuses are found in the Pacific and Atlantic Oceans at depths of 3,300 to 9,800 feet. With an average mantle length of 6 inches and arms not much longer, great big round eyes, and skin tones ranging from white to red/brown, they look like tiny, noble ghosts in the darkness. In April 2007, an American team guided an ROV to a site where they knew the deep-sea octopuses incubated their eggs, at a depth of 4,600 feet off the coast of California. The group were again successful and were able to observe a female apparently searching for an appropriate place to lay her eggs. When the team returned to exactly the same place 38 days later they were all astonished to see the same female there. They were able to identify her as the same creature because she had distinctive scars. Sitting on a slope, she was evidently guarding her clutch. Over the next 4.5 years the team returned to the brooding octopus 18 times and she didn't seem to have moved at all. While the young octopuses were developing in their capsules, their mom, over the course of the years, became thinner and paler. When the team observed the octopus nudging away an approaching crab instead of eating it, they assumed it was possible that during this long time the female had not fed at all. Even pieces of crab offered using the robotic arm on the submersible were rejected. It is generally believed that female octopuses, of all species, eat very little or nothing at all when brooding and wafting constant supplies of fresh oxygenated water over their eggs. The extremely long brooding period of 53 months did surprise the team and is the longest known for any animal. When they sent the sub down for the last time, in September 2011, there were no signs of the female and all that remained lying on the ground were the empty capsules of around 160 hatchlings. In the octopus world it is not unusual for females to die after their brood has hatched, but for one to have lived so long was extraordinary. Most octopuses living in shallower zones have a life span of 1 to 2 years, including the incubation time. The sacrifices made by octopus moms ensure the best possible start in life for the hatchlings because when they do hatch they are already well developed and are immediately able to look after themselves.

In 2016, another small deep-sea octopus captured the attention of media worldwide and became a YouTube star. During a dive of the ROV *Deep Discoverer*, the research team of the National Oceanic and Atmospheric Administration made a discovery that thrilled a global audience. At a depth of 14,000 feet, off the uninhabited volcanic Hawaiian Necker Island, a small form looked at the camera fitted to the ROV. The video was later uploaded, and the Internet community nicknamed the octopus Casper, after the friendly cartoon ghost. All alone, Casper rested on the seabed guarding her young—this Casper was female. The small octopus was of an unknown species, which made the discovery even more exciting. The anatomy of the animal resembled octopuses that were not thought to be found in the deep sea. Casper's roughly 4-inch mantle showed no signs of pigmentation: the animal was ghostly white and slightly transparent. Later the same year, another deep-sea expedition made a further exciting discovery. At a depth of over 13,000 feet, they found 2 other octopuses of the same, still unknown, species observing them while guarding their eggs. With their large black eyes and eight white arms, the cephalopods were guarding about 30 eggs stuck to the stalks of some dead sponges. These stalked sponges need manganese nodules as a foundation to settle and grow on the otherwise sludgy ground. In an experiment it was shown that sponge populations collapsed when the manganese nodules were removed, and that is bad news for the little deep-sea ghost, Casper. If there are no manganese nodules then the sponges cannot flourish there and the octopuses cannot lay their eggs there. But why would the manganese nodules disappear, you ask? Well, the manganese and other elements in the nodules are becoming increasingly important for industry, and the search for mineral resources in the deep sea has already begun.

Gold Rush in Deep Waters

The search for raw materials is driving people to ever remoter areas. Sand, gravel, oil, and gas have been extracted from the oceans for many years, but now humans are beginning to predict the possibilities of another depot of raw materials—the deep sea. Way down there, in the darkness, lie raw materials worth billions of dollars, and various countries, particularly the industrial nations, need these raw materials. The demand is high for precious metals for smartphones, cars, and solar panels. News of deep-sea deposits are in the process of starting a real gold rush. The problem is the potential harm to sea creatures by deep-sea mining is so great that the International Union for Conservation of Nature equates it to forest degradation. We simply know too little about the deep sea, and deep-sea flora and fauna could be destroyed forever by the mechanical intrusions of mining. The demand for coveted minerals, however, continues to grow rapidly. Ever-faster development cycles mean that the latest models of smartphones with ever-decreasing durability are catapulted onto the market, and broken or unwanted ones, which contain copper, cobalt, and aluminum, end up in the garbage. These phones will of course need to be replaced, and they too contain rare and precious raw materials, raw materials that can be found in the deep seas and that are waiting to be exploited. Commercial deep-sea mining is planned to start in 2025.

The Peru Basin, the site of the two previously mentioned Casper octopus broods, is home to manganese nodules. The largest, and thus economically the most important, deposits of this metal are found in the equatorial North Pacific. The almost 2-million-square-mile area stretches from Hawaii to Mexico and is often referred to as the "manganese nodule belt." This is an area slightly larger than Europe. The nodules cover roughly 60 percent of the seafloor in this belt, at depths of between 13,000 and 20,000 feet. And there they lie like potatoes in a field. Outwardly they also resemble potatoes, but they grow somewhat slower—on average roughly 0.2 to 0.4 inches every million years. The nodules form around a growth core and consist of 25 percent manganese and other pure metals such as iron, nickel, copper, and cobalt. In order to grow, they need chemical elements absorbed from the seawater or, more frequently, from the pore water of the underlying sediments.

It is these rare and very valuable metals in the 1- to 3-inch nodules that make them so attractive to the fast-growing high-tech industry. Mining licenses have to be authorized to exploit the resources. Germany has secured a section of the belt a bit larger than Bavaria. In 2006, the sector was leased for 250,000 Euros (US$275,000) by the UN for 15 years to investigate the deposits. Since 2015, Germany has had a mining license and is just waiting for the go-ahead. The prospective countries have to apply to the International Seabed Authority (ISA) for mining licenses. In the area leased by Germany, there are millions of manganese nodules lying on the sea bed at a depth of 13,000 feet. The raw materials they contain are worth many millions, if not billions, of dollars. On a purely technical level, the exploitation of the nodules is no problem—and even as early as 1978 the first nodules were transported to the surface—but the German Environment Agency assesses the consequences to the environment as "considerable." According to estimates, the mining of the deep-sea potatoes would clear an area of seabed twice the size of Manhattan (45 square miles) every year. Not only would sediment be stirred up, but the water and the seabed and all the organisms in that area would also be transported to the surface—even the tiny ghostly octopuses. The entire ecosystem would be damaged. Of course, the composition of species would change and, as the example of sponges demonstrated, species will disappear. In the deep sea, processes take a very, very long time, and when destroyed it could take decades, if not centuries, to regenerate.

138   It isn't only manganese that has caught the attention of industry. Other natural resources whet the appetite. Mineral, metal-bearing sulfur compounds known as massive sulfides can be found on black smokers. More precisely, the chimneys of the black smokers are made of sulfide ore deposits. The up to 750°F water that spouts from the vents transports these volcanogenic ores up and then back down to the seafloor. Layer by layer, they tower up to form the chimneys of the vents. While the black smokers of the Atlantic are insignificant economically (they contain only marginal amounts of valuable massive sulfide ores), the South West Pacific ones have aroused more interest. There, increased amounts of copper, zinc, and gold are precipitated, and in comparatively shallow waters of up to 6,500 feet. Additionally, a number of the areas fall within the economic zones of neighboring coastal countries—and thus are not under the auspices of the ISA—freeing these countries from having to apply for mining licenses. The plan is to start by mining metals from inactive black smokers, as these are no longer settled by the typical living communities of active hydrothermal vents. Although the technological know-how is limited at the moment, it is expanding. It is just a question of time until the deep sea is prospected extensively and on an industrial scale.

    Cobalt crusts formed on the flanks of submarine volcanoes in depths of 3,300 to 10,000 feet are also of economic interest. Deposits of manganese and very small amounts of iron, platinum, nickel, cobalt, and copper accumulate on volcanic rock, coating it over time in a crust. The quantities of trace metals found in these crusts is so small that many tons of rock have to be mined to extract relevant amounts. The unstable political situations in many of the countries in which these metals are mined, such as the Democratic Republic of the Congo, make the deep sea as a new source of raw materials seem all the more attractive. Quite apart from the major concerns about seafloor ecology, which would inevitably be affected by prospecting activities, there has been deficit in the necessary conveyor technology... up to now.

In addition to metals, a possible future energy source can also be found in the deep seas: methane hydrate. On the continental shelves around the world are rich deposits of frozen methane trapped in sea ice. It is estimated that supplies of methane from the deep sea are capable of producing twice the amount of energy as all the coal, natural gas, and oil reserves in the whole world. Of course, the exploitation of methane is of huge economic interest, but the dangers to the environment are also immense. First, we would again be dependent on another fossil energy source, and second, methane is a greenhouse gas with roughly 23 times as much global-warming potential as carbon dioxide. The justifiable worry, expressed in numerous scientific studies, is that methane gas will be released during the mining process, which would intensify the greenhouse effect. In turn, the resulting increase in water temperatures would lead to a chain reaction in which further methane is released into the atmosphere from the increasingly unstable sea ice, leading to climate collapse. At the same time, the mechanical mining could lead to instability of the underwater slopes that had been stabilized by the hydrates. The unstable slopes would then be susceptible to landslides, as in Frank Schätzing's novel The Swarm, when a disastrous tsunami swamps coastal areas of Northern Europe. Supporters of methane exploitation argue that the cavities in the ice, created when mining methane, could be filled with carbon dioxide from industry and power stations and thus stabilized. By the way, we can thank tiny methane-eating bacteria for the fact that very little methane hydrate is released by natural means; they ensure that only 2 to 4 percent of leaking methane reaches the atmosphere.

The idea of extracting suppliers of energy from the sea is not a new one. In Jules Verne's novel *Twenty Thousand Leagues Under the Sea*, published in 1870, the main character, Captain Nemo, explores the deep sea with a submarine powered by marine resources. Verne was way ahead of his time, and what was once fiction soon became reality. On board the *Nautilus*, Nemo and his crew discover all sorts of creatures in the dark depths of the sea. Even though Verne wouldn't have dreamt it, the deep down reality was, and is, far more extreme than his fiction. As incredible and fantastic as all his literary descriptions and imagery of the deep sea were, they almost always pale against what the deep sea really hides.

Sea Monsters, Deep-Sea Demons, and Sailor's Yarns

Hidden in the dark waters of the deep sea, these creatures have inspired human imagination for thousands of years—*Architeuthis dux*, the giant squid. No other mollusks are enshrouded by as much mystery as these ten-armed giants. As early as 77 BC, they were mentioned in Pliny the Elder's book *Natural History*. But its triumphal entry to world literature was in *Twenty Thousand Leagues Under the Sea*. Jules Verne describes it as a sea monster engaged in a fight to the death with Captain Nemo. Although there are plenty of stories of giant squids attacking ships, all of them have been dismissed as seaman's yarns. In 2003, the crew of the trimaran *Geronimo* had a close encounter. During a yacht race for, believe it or not, the Jules Verne Trophy (presented for the fastest circumnavigation of the world), the *Geronimo* had an unexpected visitor in the Strait of Gibraltar. A giant squid had attached itself to the upper part of the rudder and the stern of their boat. A crew member described the situation as scary when the huge creature began shaking the boat with its powerful arms. The crew estimated the total length of the squid as roughly 33 feet, with suckers about 4 inches in diameter. They slowed down in an attempt to lose their unwanted guest, and an hour after thumbing a lift the giant squid slid back to the depths. There was no photographic evidence of the incident, so people can puzzle about whether the crew took the link between Jules Verne's novel and the trophy a little too seriously or whether their version of the encounter reflects the truth. The chances of sighting a giant squid at all are extremely slim. Very few people have been lucky enough to witness a living specimen.

Most of what we now know about the animal has been gained from the remains in the stomach contents of sperm whales washed ashore. In 1856, a Danish researcher, Japetus Steenstrup, reconstructed a giant squid from the remains of one washed ashore. After more than one and a half centuries and a number of giant squid carcasses later, it is believed that female giant squids have a total length, including arms, of up to 43 feet and that the males are somewhat smaller, reaching a maximum of 33 feet. Despite its enormous size, the giant squid is only the second-largest invertebrate—the largest is *Mesony-choteuthis hamiltoni*, the colossal squid. However, we know even less about colossal squids, as they are hardly ever caught or become stranded on beaches, either of which could provide more information about them. Weighing up to 1,700 pounds, and with lengths between 40 and 46 feet, it is larger and, above all, much heavier than the giant squid. Female giant squids can weigh up to 600 pounds, and although reports have indicated that they can reach 66 feet long, they were never scientifically verified. Both mollusk species have streamlined bodies, although colossal squids look more compact and rounded.

Given our high-tech world, it is difficult to believe that the first photo of a living giant squid was only taken at the beginning of this century. On January 15, 2002, a team of Japanese researchers succeeded in catching and photographing a 13-foot-long squid off the beach of Amino-cho, in the Kyoto Prefecture. The poor beast was captured and tied to the quay in order to study it, but, sadly, it died overnight. The body was preserved and displayed at the Japanese Natural History Museum. In 2004, another Japanese team cunningly succeeded in filming a living giant squid in its natural habitat. After 2 years of preparation, they struck course for an area some 600 miles south of Tokyo. The area was well known as a hunting ground of sperm whales—the archenemies of giant squids. They theorized that the prey they were seeking could be found where the whales hunt. The team dropped a 3,000-foot fishing line baited with squids and shrimps, a camera, and a flash. After a number of failed attempts, they finally got lucky—a giant squid showed interest in the bait. The creature went for the bait and hook, and was caught for 4 hours before managing to free itself. During this time, the unfortunate squid was bombarded by flashes and science was 500 photos richer—an absolute sensation that attracted worldwide attention.

Based on the bycatch of trawlers and the stomach contents of dead sperm whales, it is speculated that giant squids usually live at depths between 1,000 and 3,300 feet, but can dive deeper. Scientists believe that colossal squids dive even deeper, maybe even to depths of over 6,500 feet. While giant squids can be found in all of the world's oceans, with the exception of tropical and polar regions, colossal squids have only been discovered in the Antarctic and the seas north of there to South America and South Africa.

These mollusks not only score with their huge bodies and deep-sea diving skills, but also hold another record. Quite a number of years ago I visited the Ozeaneum, an aquarium in Stralsund on the Baltic coast, and one of the most splendid exhibits was a preserved 20-foot-long male giant squid. What impressed me most about the creature was its enormous eyes. Actually, Architeuthis and Mesonychoteuthis have the largest eyes in the animal kingdom. The Architeuthis specimens have eyes that measure 10 inches in diameter with 3.5-inch pupils. So, the eyes are roughly the same size as a basketball or a pasta plate. It is thought that they have developed such big eyes in order to spot the silhouettes of approaching sperm whales. "Silhouettes in the depths of the deep sea? How is that possible?" I hear you say. As we have already discovered, there are many bioluminescent organisms in the deep sea. When they are disturbed, for instance by a sperm whale sneaking up on its quarry, their defense systems light up, making the silhouette of the whale visible. According to a model calculation, the huge eyes of both mollusks can see more than 400 feet and probably evolved as a result of predator pressure. This 400 feet is roughly equivalent to the range of the sonar system used by sperm whales when seeking out prey, which again shows the ingenuity of evolution in balancing out the arms race in the animal kingdom. The battle between these two giants has always fired the imagination, and curiosity is further stoked by huge imprints of suckers on the skin of sperm whales. Evidently, squids don't give up without a fight, although analysis of the stomach contents seems to indicate that the win rates are on the side of sperm whales. Sperm whales often bear the scars of these life-and-death struggles. We can only speculate about what exactly happens during these battles, but the very fact that they do actually take place is in the meantime enough to provide further materials for myths and yarns.

Another very bizarre creature of the deep sea is a relative of the wood louse you find in your cellar, but the deep-sea version is many times the size of your small cellar dweller and looks like one of the aliens in *Men in Black*. *Bathynomus giganteus*, a species of aquatic crustaceans, can weigh up to 3.7 pounds and reach a maximum length of just over 2 feet, so roughly the same size as a year-old child. French zoologist Alphonse Milne-Edwards discovered the first one of these beasts in the Gulf of Mexico in 1879. The creatures can be found at depths between 500 and 7,000 feet, but can sometimes be found considerably deeper. This giant isopod lurks on the seabed waiting for food from above to land on its head... or right in front of one of its 14 claws. Giant isopods are, first and foremost, scavengers and live off marine snow, but they won't turn up their noses at the odd morsel of whale or any other carcass, and can even actively hunt. When hunting they tend to concentrate on sessile animals such as sea cucumbers or sponges or any other organisms living on the seafloor. When filming an episode of

Shark Week for Discovery Channel in 2015, a giant isopod was captured on film attacking a spiny dogfish. The dogfish was caught in a trap and the isopod, seizing its chance, latched onto the unfortunate dogfish's face and devoured it.

Information about the isopod's habits and feeding behavior are gathered by deep-sea expeditions, but they can also be admired and studied in aquariums. The "Wonders of the Deep" gallery at the Aquarium of the Pacific in California houses four of these deep-sea dwellers. There they keep a very close track on the feeding behavior of the creatures and have discovered that their favorite meal is mackerel. Of course, in the wild the creatures are not used to luxury, as they have to live off what they can get. In aquariums, however, where they almost have to be fed by hand, they attempt to feed these creatures every day, despite the fact that they eat very irregularly. According to aquarist Dee Ann Auten, in one year—2013—one isopod ate only twice, one ate four times, one seven times, and the last one almost ten times. The aquarists in a Japanese aquarium obviously hadn't found the favorite food of their giant isopod who, after a five-year hunger strike, died on them. Normally the creatures are not picky about what they eat and gorge themselves until they can't move. Three giant isopods caught and examined in the Gulf of Mexico had ingested large amounts of plastic waste.

It seems the deeper you go into the sea, the stranger (and bigger) things become. The reason why isopods can become so large in deep water is because deeper waters are very cold and relatively well oxygenated. Add to this the fact that on a per weight basis, large animals consume less oxygen than small ones. But the cold temperature is what truly enables the isopods to grow to such large proportions: the cold temperature

reduces their oxygen requirements; it is their oxygen requirements which determine the size that water-breathing
ectotherms (also known as cold-blooded animals) can reach.

The phenomenon is called deep-sea gigantism.

When isopods are attacked, just like their terrestrial relatives, they roll into a ball to protect their vulnerable undersides.
Segments of their flattened hard back armor overlap, forming
an impenetrable protective shield. In proportion to their pale
pink bodies, their overly large compound eyes—which are
widely spaced and emphasize their otherworldly appearance—
can see as well as cats' eyes in the darkness. Another sensory
organ, their antennae, are roughly half the length of their bodies and help them feel their way forward in the darkness. Small
claws give them a bit more stability of the seafloor.

Bottom dwellers have adapted more than their anatomies
to survive in the darkness. Many of them have developed their
own specific strategies to hunt—or not to be hunted. In both
cases light proves useful.

Glitter in the Dark

Deep down in the sea it is pitch black, but this blackness is regularly interrupted by flashing, flickering, and glittering Morse
code signs. And, on closer inspection, the deep sea is not quite
as dark as it seems. We now know that as many as 76 percent of organisms there are capable of producing their own
light. Under water, however, wavelengths of light can be problematic. The blue-green range has the greatest reach with a
wavelength of around 470 nanometers, and most deep-sea animals can only pick up this range. They lack the pigments to see

other wavelengths and, unlike other bioluminescent animals on land (such as yellow-blinking glowworms), the creatures of the deep flicker a blue-green light. Of course, they don't do this just for fun—that would be a waste of energy—but instead to irritate enemies, to attract partners or food sources, or for camouflage purposes. Bioluminescence produces cold light, meaning that the animals don't overheat while producing it. Unlike artificial light sources, where about 90 percent of the energy is lost as heat and the rest converted to light, the light yield of organisms generating bioluminescence is almost 100 percent. This phenomenon is found in many sea dwellers from bacteria and algae (see the section "The Glow of the Sea"), to crustaceans and starfish, to sharks and other fish. Around 1,500 known species can glow. Light is thus the most widespread form of communication on our planet, even though on land relatively few organisms use bioluminescence. It is generated by a chemical reaction. When oxygen comes into contact with a light-emitting molecule—luciferin—energy is released in the form of light. There are different forms of luciferin that, depending on the species of animal, vary the reaction executed. For this reaction to happen in the first place, many organisms need to produce a catalyst—luciferase—to accelerate the reaction. The resulting energy is released as photons and voilà! It flickers in the dark. In principle the reaction is similar to the one used in glow sticks: a capsule is broken, allowing previously separated chemicals to mix and light up the stick. Luciferin is synthesized by some animals, while others absorb it via foodstuffs. The latter can sometimes prove problematic, as many ocean creatures are transparent. If they consume luminescent food, they could well end up as food themselves. It would be like swimming around with a neon sign in the stomach saying "Eat Me!" Here, too, Mother Nature came up with a solution: the guts of many transparent fish are not transparent—problem solved. The number of species using bioluminescence and the differences in the chemically produced reactions to make light are both evidence that forms of bioluminescence evolved independently. We currently know of 40 distinct variations.

The most common example of bioluminescence in the deep sea are the deep-sea anglerfish (Ceratiidae), who trick their prey using light. The female anglerfish do not produce the light themselves; their lodger does it for them. Light-producing bacteria live in the light organ that dangles in front of them like a baited fishing rod. The bait looks like a glowing worm, so fish are almost magnetically attracted to it. Instead of a pleasant little snack awaiting them, however, there is a mouth full of extremely sharp teeth about to bore into their flesh. The leasing relationship is fine for both parties—light in exchange for food and security. The light organ is also used to attract potential partners (what happens then is described in the "Sex and the Sea" chapter).

Like anglerfish, black dragonfish of the genus *Idiacanthus* are not only as ugly as sin, but they also use worm-like bait called a barbel. This barbel grows like a long whiskery wart from the prominent lower jaw of the female dragonfish and ends in a longitudinal elongated bulge. The up to 20-inch-long and slender body of the female black dragonfish resembles an oversized black tadpole. The creatures, found around the globe in southern subtropical and temperate oceans at depths of up to 6,500 feet, also have enormous mouths that take up most of the space of their heads and are peppered with a vast collection of dagger-like fangs. The males are one-tenth of the size and don't have any teeth, intestines, or barbel. Elsewhere, they are well equipped—their testicles take up the entire abdominal cavity. Male *Idiacanthus* evidently only live for reproduction, about which nothing specific is known. Apart from their peculiar appearance, the animals have another special feature: beneath their eyes are cells that, like the power eyes of Superman or a searchlight, can beam light. This light is not blue-green but red and, due to the color blindness of the prey, is invisible. The searchlight of the deep-sea dragonfish has a range of about 6.5 feet, giving the hunter a huge advantage. Their prey sense the ripples from an approaching predator in their lateral line organs only once the predator is a mere 8 inches away. The black dragonfish, on the other hand, have had their prey firmly in view for longer, and it is usually too late for the color-blind prey to escape.

Another really clever example of the use of bioluminescence in hunting prey is practiced by a shark that is shaped and colored like a cigar, the cookiecutter shark (*Isistius brasiliensis*). They are roughly 20 inches long and are at home in all oceans but prefer tropical and subtropical regions. During the day they are usually found at depths of between 3,250 and 13,000 feet, and at night they swim upward, mostly remaining about 300 feet below the surface. These small sharks like to eat squids, but what they like most of all is biting bits off larger sea dwellers. Sea mammals such as sperm whales, dolphins, humpback whales, seals, rays, and sharks big and small, as well as tunas and other feared marine predators have all been found with the typical bite marks. In order to get their meal, the cookiecutters use a trick. Their paler underbellies are covered in photophores—luminescing light organs—except for a darker collar around the throats and gills. When a larger predator (that is, their prey) approaches from below, the bioluminescing silhouette merges with the backlight (in this case, the moonlight) using a strategy known as counter-illumination (see too the "Perfectly Hidden" section). While almost the entire body is now light shaded, the collar around the throat remains dark colored. From below their silhouette now looks like a small fish, which will, in all likelihood, attract a predator. The cookiecutters probably use this decoy to satisfy their own hunger, in which case they are the only organisms to take advantage of the absence of bioluminescence (in the neck area) to attract prey, with the photophores contributing to camouflage. The teeth of cookiecutter sharks are also remarkable, as they leave recognizable imprints in the bodies of their prey. The lower teeth are triangular in shape and aligned in one single row, while the upper teeth are more pointed and, in relation to their bodies, the longest of all sharks. A bite from a cookiecutter shark looks similar to an oval cut from a piece of dough, hence the name. These small brown sharks are pesky little critters, and there have been reports of whales washed ashore with dozens to hundreds of cookiecutter bite marks. Bite marks—old and new—on the bodies of almost all spinner dolphins off the coast of Hawaii are evidence that they have become acquainted with the creatures. A couple of people also bear the scars of a cookiecutter attack, and although the inch-long wounds are painful, they are not life-threatening. Encounters with these sharks, however, are extremely rare, as they are not usually found near the coast. There have been reports of the little beasts attacking offshore divers or of shipwrecked sailors being nibbled by the animals at night. Apparently cookiecutters are not particularly choosy about what they eat, as they have also tried bites out of submarines and underwater telecommunication cables. Actually, "vampire sharks" might be a better name for these little omnivores.

Speaking of vampires, another spectacular glowing deep-sea denizen is *Vampyroteuthis infernalis*—literally translated as "vampire squid from Hell." The skin webbing stretched between their eight arms is to thank for the name, because it resembles a cloak often associated with vampires. The webbing functions as a kind of sail, which enables the cephalopods to move effortlessly through the depths, usually between 2,000 and 3,000 feet. Vampire squids have another two tactile velar filaments that can be extended to enormous lengths. These strange creatures feature two blue or red eyes that take up almost a sixth of their total body length. The color of their up to foot-long body varies and, depending on the place and the light conditions, appears anything from a velvety black to bright red. Almost their whole body is covered in photophores that emit flashes of light, the intensity and duration of which is regulated by the squid. Vampire squids, when threatened, eject blue bioluminescent clouds that glow for up to 10 minutes and are intended to confuse the attacker and give the squid time to flee in the darkness. Mature adults navigate and move by using the two little fins that stick out of the sides of the mantle and look like mini elephant ears. Even though the name suggests an unquenchable thirst for blood, the squids feed on detritus—the marine snow—which, using their suckers, they cover in slime and then consume.

One of the world's most remarkable creatures doesn't produce light itself but benefits from sea creatures that do. The barreleye or spook fish lives in the mesopelagic zone at depths of up to 3,250 feet. Very little light penetrates this deep, but this weird fish has adapted to its dark surroundings in a way that is as ingenious as it is crazy. They have transparent gelatinous heads that are extremely fragile. If you were to keep the fish in an aquarium, the first little bump against the glass wall would be lethal. As of yet, only one single time has somebody succeeded in capturing a barreleye and keeping it for a few hours in an aquarium before it died. Researchers at the Monterey Bay Aquarium Research Institute were able, in these few hours, to solve the puzzle of its eyes. The eyes lie not on the outside of the head but within it, each in a green-blue dome. In order to maintain a perspective on things, barreleyes, although generally gazing upward, are able to rotate their eyes within the dome. They look a little like the funny plastic heads with rotating eyes that we used to buy for a few cents in gumball machines (for those who remember them; for others, think of the magic eyes of the auror, Alastor "Mad-Eye" Moody, in the *Harry Potter* books). This mere 3-inch-long fish drifts motionless through the water, eyes gazing upward, on the lookout for potential prey like little glowing beasties. In doing so they look pretty miserable, as if mulling over the world's problems. Its round face is not unlike a child's sketch, except that the two black circles above the spiky mouth are nostrils and not eyes. It must be the only fish that manages to express melancholia with its nostrils. These animals are extremely rare, and only a few research teams have succeeded in finding them. Luckily, a few photos and even some videos of these bizarre creatures exist on the Internet, and even if they look like a bad photomontage, they give us a realistic impression of these special animals.

Fragile Carnivores and Reefs of Glass

Sponges, we know, are by far not the simple organisms that they seem on first glance. Being tied to one spot, they have developed an arsenal of effective weaponry to ward off hungry hunters. We find them from the shallow zones to the deep sea, where they are the lords of the benthos (the entirety of organisms found on the seafloor). As with all deep-sea denizens, we have happened upon their secrets bit by bit. Many deep-sea sponge species are suspension feeders, filtering bacteria from the water. Some species, however, prefer food with a bit more substance: carnivorous sponges. Yes, there really are sponges that have nothing to do with your bathroom sponge. Early this millennium, off the Californian coast at depths between 10,800 and 11,500 feet, a carnivorous sponge was discovered that neither looked like nor behaved like a conventional sponge. *Chondrocladia lyra*, as the Latin name suggests, resembles a lyre or harp or an upside-down wide-tooth comb. Rhizoids anchor these sponges to the soft sediment of the seafloor, and between 1 and 6 horizontal stolons spread out star-like from a central point, with roughly 20 branches rising bolt upright from them. These branches are equipped with barbed hooks and spines that form an impenetrable net. Each branch ends in a small bobble filled with spermatophores. When these are released in water, they might, with a lot of luck, come across another hermaphroditic lyre sponge and fertilize it. The egg cells of the sponge are located at the central point of the star. Should a copepod, or another crustacean drifting by on the current, become ensnared in the net, its fate is sealed: the sponge secretes a digestive membrane breaking down its prey, which will be absorbed later. The lyre sponges show, in an unusual way, how animals can adapt to darkness and conditions with very few resources and still prosper.

From a deep-sea musical instrument, we now turn to deep-sea lighting. Another carnivorous sponge (there are currently some 33 described species), looks no less peculiar. The predatory, long-stalked sponge *Chondrocladia lampadiglobus*, also known as the ping-pong tree sponge, looks astonishingly similar to an art deco lamp, with 10 to 20 glass globes on stems radiating from the top of a single stalk. These lamp-like, milky-colored globules house the gametes. The central stalk of the lamp, the rhizoid, anchors the sponge to the seafloor. This spongy work of art prefers anchoring in the Eastern Pacific at depths between 8,500 and 9,800 feet. Ping-pong tree sponges are around 20 inches tall, a bit more than half of which is the thin stalk. The pretty sponge catches its prey using sharp, hook-shaped spicules that cover its body but in particular the exterior of the globules. Crustaceans drift by on deep-sea currents and are trapped on the sharp spikes.

Sponges contribute to the diversity of species by providing preferred brooding grounds (see the "Life at the Extremes" section) or a habitat for other organisms. A very fascinating example is a sponge that excites both architects and engineers. The white Venus's flower basket, *Euplectella aspergillum*, has a skeleton consisting almost entirely of biological silica (an oxide of silicon). Like corals, this living organism produces an inorganic substance whose building blocks are extracted from seawater. The whole construction of the sponge resembles an 8-inch-long and 1.5-inch-thick baseball bat that has been firmly anchored in the muddy sediment by needles, the spicules. The deep-sea club is, of course, not made of wood but of a densely entwined network of glass lamella fibers of varying thicknesses and spiral-shaped ribs. The skeletal structure of the sponge is incredibly complex, featuring at least seven levels of the glass fibers in thicknesses ranging from microscopic to an inch or so. The silicate or glass fibers are special in their own right, as they have the same properties as commercial fiber optics made of plastic or glass, and in some cases are even better conductors of light. The sponge fibers are also considerably more flexible and significantly more stable than technically produced fibers. And sponges have to be stable because the pressure in the depths of the ocean is very great and they have to endure strong currents. The arrangement of the fibers of the Venus's flower basket sponge was already being used in the nineteenth century as a textbook example for architectural constructions like, for example, the Eiffel Tower. The sponge was known at that time because it was found at depths of only 130 feet. Why the sponge needs light-conducting fibers remains a mystery.

Sponges are essential for life in the sea: they filter huge amounts of water as well as provide a habitat for a vast array of creatures. With its lattice structure, the Venus's flower basket is well suited as a shelter, and spongicolid shrimp larvae are well aware of this. These shrimp larvae swim, mostly as pairs, through the latticework of the sponges, one developing to a male and the other to a female. Afterward, something happens that often occurs with couples: they stay at home, make themselves comfortable, and dine together. After a while, the shrimps grow to such a size that they can no longer fit through the lattice—they are imprisoned in a glass cage. Actually, it is a pretty fine arrangement for both parties, sponge and shrimp. The latter do the indoor chores, cleaning and getting fresh water and food, and in the security of their prison produce plenty of offspring, who can pass through the lattice structures to the outside world. This sponge is a popular wedding present in Japan, due to the symbolism of the lifelong, albeit unintended, partnership of the shrimps.

Glass sponges are not really as rare as you would think; there are even whole reefs of them. Off the coast of British Columbia, the westernmost province in Canada, there is a 10,700-square-foot glass sponge reef, although not a deep-sea one. This reef was discovered in 1987 by a team of Canadian researchers at depths between 500 and 800 feet. The discovery of this 9,000-year-old reef was a scientific sensation—it was previously thought that glass sponges had become extinct 40 million years earlier. Glass sponges have been around for some 545 million years, and had their heyday during the Jurassic period when dinosaurs roamed the Earth. At that time, huge glass sponge reefs flourished in the prehistoric seas. Today all that remains are the gigantic limestone rock formations stretching from the Caucasus—via Poland, Germany, Switzerland, France, Spain, and Portugal—all the way to Newfoundland and then Tennessee in the United States. The reef off British Columbia, however, lives and flourishes, with cup-and funnel-shaped sponges next to others resembling flowers. In fact, some structures there tower as high as eight-story buildings, with glass apartments offering refuge to the various animals seeking shelter.

These sponge reefs, like the coral reefs, are nurseries for many fish species and are important to the marine food chain. The glass sponge reefs, found only there and off the coast of Alaska, are so fragile that sediment disturbed by fishing boats could damage or even kill the reefs. Furthermore, bottom-trawling nets leave a trail of destruction on the seabed behind them, plowing up everything in their way. Sadly, this is already the case and a large portion of the reef has already been destroyed by the trawling nets of the fishing fleets.

Glass sponges, regardless of whether they live in the reefs or individually, are threatened because they react very sensitively to climate-induced temperature fluctuations and can only live in a narrow range of salinity. They also need plenty of dissolved silica in the water for growth. Conservation agencies and specialist scientists are working hard to protect this remnant of the dinosaur age because what took 9,000 years to grow could be completely destroyed in a couple of hours. Just recently it was discovered that glass sponges also grow on manganese nodules. And, sure enough, sponges are among the most common benthic macrofauna in the manganese belt. If the manganese nodules are mined industrially, it could take decades or even centuries for the glass sponge populations to recover. We have to know a lot more about the various ecological systems in the deep seas before we start brutal intrusions, otherwise these major interventions might have unexpected consequences for all of us.

SEX AND
THE SEA

~~~~~~~~~~

S EX ENSURES THE continuation of the species and, as we
have discovered, there are a number of peculiar variations in the sea—just think of Nemo! For humans, it can
be somewhat unsettling to discover that the marine animal
we have always considered cute and fuzzy is actually a double-
dealing kidnapper and rapist. Penis fights, romantic castle
building, tentacle petting, and fish that swim thousands of
miles to "do it" for the first and last time are just a few examples of what we discover when we look closer at the most
important subject in the world.

The Black Soul of the Sea Otter

~~~~~~~~~~

Who doesn't know these cuddly toy-like animals of the sea:
black button eyes, snub nose, fluffy fur, adorable paws... Sea
otters (*Enhydra lutris*) are the very definition of adorable. The
female looks so cute, drifting on her back on the surface of the

sea, devoting herself to grooming the fur of the fluffy pup resting on her stomach. Sea otters are also well known for their use of tools, indicating a high degree of intelligence. In the course of evolution, the animals have learned that difficult-to-open mussels break if you hammer away at them long enough with a stone. Other animals with shells, such as crustaceans, sea urchins, and clams, suffer similar treatment. In addition to stones, sea otters also hit their next meal against other hard objects like ships' hulls or bottles lying on the seafloor. In doing so, sea otters join the crème de la crème of intelligent animals. The use of tools is confined to very few other known animals, such as crows, dolphins, apes, and more recently, *Choerodon schoenleinii*, the blackspot tuskfish, who smash mussels against rocky slopes before eating them. When sea otters find a particular tool—for instance, a small pebble—has proven to be especially effective or sits well in the paw, it is hidden in a fold in their fur and kept for future use. The coat of the otter is very special and, unfortunately, the reason that sea otters were hunted almost to extinction between the eighteenth and twentieth centuries. Sea otters, unlike other marine mammals, don't have a protective layer of blubber; they are completely reliant on their fur for warmth. Sea otters have the most dense and finest fur in the animal kingdom, with nearly 1 million hairs covering an area of only 1 square inch. The fur consists of coarser water-repelling outer hairs as well as a layer of underfur that provides perfect protection against the cold while also helping with buoyancy. To be more precise, the tiniest of bubbles blown by the otters into their fur provide both insulation and buoyancy as well as increasing lung capacity, which also helps buoyancy. But the clever tool users can do more. They take advantage of the stability and tensile strength of kelp to prevent nimble prey like crabs from escaping: they simply tie up their future meal with kelp. When they feel like a bit of kelp wrap, they return to untie their meal, crack the shell, and tuck into the gooey bits. This strategy is also used by sea otter moms to prevent their pups from drifting away while they hunt for food, not unlike human parents who tie their kids to the buggy to stop them from vanishing in the supermarket.

But these cute little workers have their dark side. Sea otters may well be one of the most adored creatures on the planet, but beneath their silky fur hides a sinister character—well, at least beneath the fur of the males. Male otters are very territorial and seek out areas with a high proportion of females to mate. And their mating rituals are very rough: the females often receive a bloody nose or suffer even worse: death. During the act, which takes place stomach to stomach, the male bites the face and nose of the female as anchorage in an attempt to stop slipping off her silky fur. This causes injuries to the nose, which is sometimes even bitten off. On top of this, during mating season the male sea otters harass the females relentlessly. When a female is caught, she fiercely resists the raw advances of the male. In an attempt to make the female more compliant, the male sea otter will often force the female's head under water, and sometimes they do not survive this brutal act. Even after death, the male sea otter will still try to mate with the drowned female. But this is just the beginning. Males who don't have their own harem search for other helpless victims in the neighborhood and choose the easiest of all—seal cubs. The juvenile seals are treated just as roughly, the ordeal of forcible copulation often ending in the death of the seal, which doesn't prevent the otters from continuing to mate. There have been reports of an otter abusing the decomposing body of a juvenile seal for up to a week after its death, thus totally destroying any semblance of its cute marine mammal image. And if rape and murder weren't enough, then there is always kidnapping and extortion. When food is short, the male sea otters kidnap the cubs of female sea otters and hold them hostage until the females come up with the ransom—in this case, food. Silky fur or no silky fur, male sea otters are among the nastiest creatures on the planet.

Sodom and Gomorrah at the South Pole

Of course, in addition to cute sea otters (at least at first impression), there are other crowd-pullers in the animal kingdom. For instance, penguins are adorable fellows in black tuxedos, jauntily tottering around and sounding like broken vuvuzelas. In the penguin world everything seems to be dandy, and the animals live pretty conservative existences: monogamous and nest-to-nest like millions of people everywhere. But looks can be deceiving, at least among Adélie penguins (Pygoscelis adeliae). Every year in summer, thousands of these cute knee-high penguins congregate in Antarctica to reproduce and raise their chicks. To do so they form pairs, often with last year's partner, together building a stone nest. In Antarctica, good stones are difficult to come by, and thus are hard currency for the tux tribe. As with all sought-after goods, a brisk market becomes established, but in secret, underhandedly, or rather under wing. When temperatures begin to rise toward the end of the breeding season, the melting ice means danger for the eggs and nests have to be shored up by small pebbles. The females set off in search of suitable nesting materials. In order to avoid the inevitable conflicts that occur when stealing stones from the neighbor's nest, the females waddle off to the periphery of the penguin colony. This is where the bachelor groups of male penguins live, the ones who hadn't succeeded in finding a partner. Those without wife and offspring have plenty of time to amass riches—namely, pebbles—and some of the bachelors build proper cairns out of them. But, of course, not having a partner also has disadvantages: the unwanted males are bursting with pent up sexual energy and want to spread their seed. The clever girls realize this and make explicit offers to the cairn-builders by flirtingly signaling a willingness to mate. Once the act is completed, the female snatches a pebble as payment for her efforts, waddles back to her unsuspecting partner who has been patiently waiting at home incubating the egg. Prostitution, the oldest profession, is well established down there and common practice among penguins. However, during this transaction—sex for stones—things do not always stay above board. Females have been observed making such elaborate courtship rituals, and bewitching the young males to such an extent, that the males became careless or too slow. The females then snatch a stone, without providing sex in return, and set off posthaste toward their nests. One crafty old hand was so accomplished at this trick that she managed to scamper off with sixty-two valuable stones!

However, Antarctic explorer George Murray Levick stumbled across the darkest secrets of Adélie penguins early in the twentieth century. What he observed was so shocking to him that he wrote part of his report in classical Greek, just in case his notes fell into the wrong hands. He had decided to make his notes available to only to a small elite of educated, upper-class males who were able to understand classical Greek. His notes were then rediscovered decades later and finally published in 2012. According to Levick's detailed account, penguin perversions range from masturbation, sexual coercion, and rape to necrophilia. The explorer found the following observation particularly galling: an injured female could only move by crawling on her stomach and was in such a state that Levick considered putting her out of her misery. One of the males named by him in his account took advantage of the opportunity and forced himself upon the female. Levick noted: "There seems to be no crime too low for these penguins." In all fairness, it should be pointed out that Levick had quite possibly misread the situation. A female lying on her stomach signalizes her willingness to mate; the male could quite easily have misinterpreted the wounded female's behavior and acted accordingly. But Levick also observed male penguins mating dead and frozen animals of the same species, as well as the sexual harassment of young chicks. There were often consequences for the juveniles, some of which were fatal. Sodom and Gomorrah at the South Pole!

Sometimes Size Does Matter

They are everywhere, although not particularly obvious as they are so small. You are most likely to know them as growths on hard surfaces like rocks and ships' bottoms, but they also grow on living creatures like humpback whales, turtles, and clams. On animals they look like light-colored warts from which hairs sprout. Okay, that comparison isn't particularly charming and is unfair to these fascinating creatures. The pure marine living barnacles (Cirripedia) belong to the Crustacea subphylum.

Of roughly 120 species, some are sessile and others parasitic. A very well-known and widespread representative of this type is the acorn barnacle (*Semibalanus balanoides*). In Germany, the poor beasts really don't have it easy, partly because their local name—common seapox—isn't likely to win over many hearts. But if the sea otters and penguins have taught us something, it's that sometimes it is better to avoid making premature judgments based on external appearances. These little "pox" start their lives as larvae, passing through two stages: The nauplius stage develops after a fertilized egg, incubated for about 6 months by one of the parents, hatches. The cypris larva stage begins after a series of molts and an intensive phase of looking for a suitable hard substrate—such as the shell of another animal, the hull of a ship, rock faces of the intertidal zones, jetties, mooring piles, and so on—preferring places where other barnacles have already settled. Once they have found a good spot, the acorn barnacles cement themselves firmly to the substrate and begin the metamorphosis to adult animals. Afterward the mini-barnacles begin building their own fortresses, a conical calcite shell enclosed by another two-part lid. Unlike other barnacles, the base of the shell of acorn barnacles is not calcified but membranous. The shell consists of six small calcified wall plates that, like all crustaceans, are molted during growth. For protection against dehydration and natural enemies, they can close their little fortress with two small lids. Acorn barnacles have many enemies, from snails and crabs to sea urchins and birds. To enable them to eat efficiently, the acorn barnacle's legs open up fan-like to form combs, peeping out from the peaks of the shell to filter plankton and other detritus from the water. Most barnacles are hermaphrodites, having both male and female sexual organs. Like almost all crustaceans,

they reproduce internally (just to compare: most corals fertilize externally, releasing gametes into the surrounding waters).
Like corals, however, the sessile barnacles have a problem
when reproducing. Since they can't get up and go off in search
of a partner, Mother Nature had to come up with something
special. Acorn barnacles have enormously long penises—and
when I say long, I mean really long. The penis is, believe it or
not, eight times the length of the body of the little barnacles,
making them the animals with the greatest body size–penis
length ratio in the animal world. The famous scientist Charles
Darwin was so fascinated by this that be became seriously
obsessive about the organisms, particularly their reproductive apparatus. He studied the small barnacles and their male
sexual organs, which he described as "prosciformed," for all
of 7 years. The sexual organ of the barnacle does indeed look
a little like an elephant's trunk, but some concessions have
been made to adapt to different conditions. Animals living in
rough waters tend to have smaller, more compact and muscu-
lar penises, while animals living in calmer waters have longer,
thinner, and more flexible penises. When searching for a partner the barnacles unleash their trunk-like penises to feel out
the surroundings in the hope of finding a mate.

Other crustaceans are as round as peas or kidney-shaped
like broad beans (but much smaller). Ostracods, sometimes
known as seed shrimps, reach a maximum size of up to 1 inch,
but on average are much smaller, maybe 0.02 to 0.08 inches.
They can be found in all aquatic environments, from groundwater and rivers to lakes and the deep sea. Bodies of ostracods
are protected by two calcareous valves that split their bodies
symmetrically. Their diet, depending on species, is varied and
includes almost everything: there are filter feeders, scavengers;

and grazers that eat algae or suck away at plants; others eat meat, arrow worms, or hunt tiny young fish. Unlike most of the sessile bivalves or copepods, the ostracods can crawl on the seafloor, swimming occasionally. These creatures need plenty of space for reproduction—literally. Their reproductive organs take up about one-third of their total body volume. The organisms are gonochoric, meaning that one individual has a male reproductive organ and another has a female one, so while the males have two sperm pumps, which take up most of their body volume, the females have two long channels to the two vaginal openings. There is a reason that the reproductive organ is so extensive: by far the most widespread form of reproduction in the animal kingdom consists of producing as many sperm cells as possible while expending as little energy as possible. Ostracods, however, throw this tactic overboard. They are interested in quality not quantity. The minute creatures (typically around 0.04 inches) produce sperm cells that are up to ten times the size of their own bodies. Putting that in human terms, a 5-foot-10-inch male would have to produce 510-foot sperm to equal an ostracod. The tiny crustaceans with huge sperms have evidently found a certain amount of success with their strategy, as their reproductive organs have hardly changed since the Cretaceous period 100 million years ago, when dinosaurs dominated on land. After mating a female barnacle, the sperm from the males equipped with two penises is stored in special sperm receptors until oviparity (the production of eggs). The females don't store sperm from single individuals but mate with a number of individuals and store the sperm. So, the longer the sperm, the more likely a favorable outcome for the males who have invested much energy in the development of their huge sperm. The successful sperm is the one that, on account of its length, manages to win the battle of fertilization. So, at least as far as ostracods are concerned, size is the key that they have used for hundreds of millions of years to triumph over other less successful and now extinct species.

Nature really is astonishing, supplying the smallest creatures with massive penises, proportional to body size, and then equipping an even smaller animal with huge sperm. But, as far as sexual organs go, the larger animals on our planet, of course, are also impressive. Bull blue whales have the largest sexual organs of all in the world. The penis of an adult male can be all of 8 to 10 feet long, with an average width of 1 foot, and weighing between 370 and 1,000 pounds. On average, the length of penis corresponds to about a tenth of the whole body length. Each of the two testicles can weigh up to 150 pounds—roughly the equivalent weight of an entire human. Their hearts are about the size of a VW Beetle and weigh 4,500 pounds. Once a bull blue whale has conquered the almost as big heart of his belle, there is not much time for romancing—competitors are lurking. During intercourse, the bulls ejaculate almost a pint of sperm. This volume both increases the chances of successful fertilization and washes away the sperm of previous suitors, as female blue whales are well capable of having a number of partners. If the mating was fruitful, then a calf is born within 10 to 12 months. Apropos growth rates of early development, here too blue whales are record holders: the embryos are the fastest growing of all mammals, and at birth a calf is a good 20 feet long and weighs 2.5 tons. The calf is suckled for some 7 months and drinks about 1,000 pints of milk a day. Once weaning is complete, the calf has grown to almost twice its birth size.

But while we are on reproductive appendages, blue whales are not the only whales with superlatives. Compared to the testicles of the southern right whales (*Eubalaena australis*), blue whale testicles are mere peas! With each pair weighing around a ton, the southern rights have, by far, the largest and heaviest testicles in all the animal kingdom. In relative terms, their testicles are 4 times the size of the apparently diminutive balls of the bull blue whales.

The genus of baleen whales (*Eubalaena*) consists of three species: the North Atlantic right whale (*E. glacialis*), the North Pacific right whale (*E. japonica*), and the southern right whale (*E. australis*). The poor old right whales get their name from being the "right" target for whalers to hunt, as they are slow swimmers and, on account of their high fat content, float on the surface when dead. Up to the middle of the twentieth century, whalers hunted these creatures almost to extinction. The numbers have never recovered, and today they are still categorized as threatened.

Of Swords and Free-Swimming Tentacles

Along with mating, penises are also used as weapons. Some of the hermaphroditic flatworms living in coral reefs have proper fights using their male sexual organs. This is called penis fencing because the hermaphroditic creatures cross "swords" as a prelude to mating. Flatworms look very similar to ordinary worms that evolution has reworked with a rolling pin—so somewhat flat. Depending on species, they can be extremely colorful or monochrome. Sometimes they are mistaken for nudibranchs. When free-living flatworms (there are also parasitic species) move through the water with wave-like, graceful motions, they look a little like a flying carpet experiencing turbulence.

When the time is ripe for producing little flatworms, the squashed hermaphrodites have to decide whether they are to be the male or female part of the equation. Since it costs a lot of time and energy to be the female and to bear the off-spring, the creatures do battle for the right to inseminate the other worm and not to be inseminated. So, they pump them-selves up and cross penises. The fencing flatworms, each with two penises, try to get rid of their sperm by stabbing the other with their penis swords. The Persian carpet flatworms, *Pseudo-biceros bedfordi*, species don't inseminate internally but ejaculate on the exterior of the opposing flatworm. The ejaculate causes the skin to dissolve, helping the sperm find its way to the eggs to fertilize them. In the process it is possible that the insem-inated animal receives lasting injuries. After the exhausting sword fight, which can last up to an hour, the loser sets off on a hunt for food to build up resources for the development of the offspring. If there are no other flatworms, near or far, then the lonely worm seeks out drastic solutions to ensure flatworms for the future. When the flatworms *Macrostomum hystrix* are desperate, they self-inseminate. Okay, self-fertilization is not that unusual in nature, but it is in this case: free-living *M. hystrix* drill their stiletto-like penises into their own heads! The sperm then migrates through the body to the egg cells and fertilizes them.

Of all the numerous crazy reproduction practices, one in particular stands out. *Argonauta*, also called paper nautiluses, are related to octopuses and, like them, have eight arms. They live in the open oceans, remaining mostly in the epipelagic zone (the sunlit uppermost part). They can be found, even though seldomly, throughout the world in tropical and sub-tropical waters. The name "paper nautiluses" refers to the paper-thin brood chamber, which is hidden by the females mostly for the protection of eggs but is also used to maintain buoyancy by pumping in surface air. The paper nautiluses are gonochoric and exhibit distinct sexual dimorphism. This means that males, with their body lengths of less than an inch, are many times smaller than the 4-inch females, whose mantles can sometimes even be as long as a foot. The males also have much shorter lives and probably die after mating for the first time. Females, on the other hand, are capable of bearing offspring any number of times during their lives. The fascinating thing about the mating rituals of the argonauts is that it takes place without the males or, rather, without their full bodily presence. Like all cephalopods, the paper nautilus has a special sexual organ—the hectocotylus—a tentacle only used for inseminating the female. The male transfers sper-matophores, small sperm capsules, to the mantle cavity of the female where the eggs released from the fallopian tubes are to be fertilized. Unlike the other cephalopods, however, the fer-tilization by paper nautiluses takes place without direct body contact. The sperm-laden mating arm detaches itself from the remaining male body and swims off in search of a female to fertilize!

You may recall from the earlier discussion of squids that the sperm of cephalopods are enclosed in small capsules. The spermatophores are not only the packaging but also the tool to implant the sperm. And they have proven to be very suc-cessful, as a sixty-three-year-old Korean woman discovered in 2008. In Korea, raw or semi-raw squids are very popular, and the lady from Seoul was very fond of them. She prepared

a live Japanese flying squid (*Todarodes pacificus*) for her evening
meal by parboiling the whole creature, including the innards.
Unfortunately, not for long enough. After briefly chewing the
very first bite, she felt a stabbing pain. She spat out the contents but still had the impression that a large number of creepy
crawlies were squirming around in her mouth. This prompted
an immediate visit to the nearest hospital. There the doctors
found twelve tiny, white spindle-shaped organisms that had
dug themselves into her tongue and gums. These "organisms"
were, in fact, the squid's spermatophores. As she had under-
cooked her meal, the spermatophores were raring to go and get
on with their job. In this case, however, instead of inseminating
a suitable female squid they chose the lady's mouth. Of course,
the doctors immediately removed the sperm capsules without
further harm to the patient, but I doubt she will ever eat squid
again. In Japan, where they also enjoy eating raw squid, a number of cases of mouth insemination have also been reported.

Sacrificial Fathers

In 1995, something odd was discovered under water. A few hundred miles off the coast of Japan divers noticed strange, circular
structures on the sandy seafloor. In 2013, after years of puzzling
about these symmetrical structures reminiscent of mysterious crop circles, scientists finally found the solution. The
architects of these sand-sculpted masterpieces are the roughly
5-inch-long male pufferfish of the genus *Torquigener*, who create these artistic formations to impress and attract females
willing to mate. The sandcastles showcase the endurance and
strength of the males, and serve as nests for their progeny.

Building a nest with a diameter of some 6.5 feet requires the male to spend 7 to 9 days working nonstop. First, they plow little furrows in the seabed that radiate from a central point by grazing the sandy floor with their bellies and vigorously flapping their fins to brush aside the sand. In this way, small sand mounds are formed along the individual radial furrows emanating from the circle's middle point. The completed masterpiece looks like a sun with wave-like rays in all directions. Once the foundations are complete, the fish decorate the mounds with coral fragments and snail and mussel shells. Naturally, the construction of such a work of art is a race against the currents and waves. The little creatures are exhausted afterward, but if the little guy, against all odds, is able to build such a beautiful nest, then it is better than any aphrodisiac. It is the living proof to the female that this male has strength and staying power—so, the ideal partner for her babies.

Only when the decorations are complete do the females begin to show interest. They thoroughly inspect the nest and are very picky. If the sand in the center of the artwork is too coarse, she just swims off, ignoring the days of toil that went into its creation. The ladies are only interested in very fine sand. Then they will deposit their eggs in the center of the nest and disappear, leaving the fertilizing and care of the eggs to the males for the next 6 days. During this time, while the father is devoted to looking after the offspring, the artwork begins to disintegrate. Once the juveniles have hatched, the male sets off in search for a suitable site for a new work of art and begins again.

Another impressive form of sacrifice is displayed by the male of a species that we have already come across—the deep-sea anglerfish. Actually, we know very little about these deep-sea monsters, and most of the conclusions are drawn from dead animals in jars of formaldehyde occupying dusty shelves in museums and universities. For example, there are only 14 specimens of the fantail angler (*Caulophryne jordani*) preserved in alcohol in natural history museums. As all these specimens were females, people were rightly asking what had happened to the males. The truth is, they had never been reported let alone caught—until recently. In March 2018, *Science* published photos that caused a flurry of excitement in the world of research. Deep-sea researchers Kirsten and Joachim Jakobsen succeeded in filming anglerfish at a depth of 2,600 feet off the Azores. During a five-hour dive in a research submersible, along a steep wall off the island of São Jorge, an unusual creature attracted their attention. Using their integrated video camera, they recorded a video that they could later evaluate when they returned to the surface. The footage revealed something astonishing: a fantail anglerfish or, to be more precise, a 6-inch female, drifting in the darkness. Typical of anglerfish, the fantail also has an angling rod (illicium) with a cold blue bioluminescent organ (esca) as a lure. It wasn't just the lure that was shimmering, however, but also the thread-like filaments spreading out from the animal's body, giving the researchers a proper light show. As if this weren't enough, on the underbelly of the potato-shaped fish they spotted an appendage an inch or so long with its own tail fin. The researchers had solved the riddle of the missing males. What they had seen was a male mating with the much larger female. It was sensational. Never before had a male been filmed while mating. The act of copulation among anglerfish is, to say the least, very odd, as the males literally dock onto and then fuse with their chosen partner. It is speculated that the males are attracted to the females by pheromones, with the enticement of the lure an added possibility. Once the female has been found, the male bites into the skin and connects to the circulatory system of the host. From then on, they are as one individual. The parasitic tiny male is completely dependent on the host and is fed via her bloodstream. The male gives up any semblance of his own existence and simply becomes a sperm supplier solely responsible for fertilizing the eggs of his host. When the female dies, he dies, too. Some species of anglerfish (caught or dead specimens) have been found with more than just a single male—the record being eight males fused onto one female. Utter selflessness in the service of propagation.

Mother Nature hasn't made it particularly easy for another marine species when it comes to the preservation of its own kind. For certain members of the Decapoda order—10-footers like lobsters (Homarus)—something crucial stands in the way of reproduction: namely, the shell. Most people only know the lobster as a delicacy. What few people realize is that before their meal was dropped alive into boiling water, the lobster was a considerate lover who touchingly took care of his partner. The nocturnal and solitary creatures are thoroughly domesticated, building their accommodations in cracks in the rocks, in piles of stones or burrows. These dwellings are meant to offer protection because, particularly when they shed their shells, they are extremely vulnerable to predators. When growing, lobsters have to shed their external shell, the exoskeleton. In the first year of their life this can happen up to 44 times. Before molting, a new layer of skin forms beneath the old armor that initially is tender and vulnerable. The lobster continues to grow during the time in which the new skin needs to become hard—for adult lobsters the hardening process can take up to a month. They spend this time in the shelter of their burrows or cracks. In the breeding season, which for both the European lobsters (*Homarus gammarus*) and the American lobsters (*Homarus americanus*) is late summer to fall, the females visit the males in their dwellings. They don't choose any old lobster but prefer males that demonstrate dominance over the other male lobsters in the vicinity. In a display of readiness to mate, the female directs a jet of urine toward the male with her head. That's right! The urine is squirted out of nozzles on the side of the head. Afterward, the pair don't get down to business straight away but first get to know each other a bit better. This closer inspection using their antennae can last a couple of days. The antennae are equipped with very sensitive chemoreceptors that help in precisely analyzing the messages in the pee. If the chemistry is right, the female accompanies the male to his lair for further declarations of affection. After a couple of days of communal living, the act begins with the female molting her hard shell, which is later eaten by the male. The female is now soft and vulnerable, and needs some time for the new shell to harden enough for her to be mobile. Once she is ready, the male carefully turns her on her back and insemination takes place belly to belly. The male transfers his packet of sperm into her sperm pouch for fertilization. Hatching then occurs the following summer. The act itself takes just 5 minutes. As the armor plating needs a couple more days to completely harden, the female remains in the male's lair. The whole time they are together, the male takes care of his partner's needs and protects her from predators and prying males. Once the shell has hardened, the female leaves the male's domain and returns to her own habitat. Shortly after the female has left, the next female willing to mate, who had been waiting patiently outside for her turn, moves in. The exiting female uses the sperm packet to fertilize the eggs, which she carries in her tail until they hatch as larvae the following summer. She carries up to 100,000 eggs, making sure they are supplied with fresh water, aired, and cleansed by fanning her tail. The small larvae, of which only 0.005 percent will survive (so, from the whole clutch only 5), live as plankton, molting three times during this stage. Afterward, they seek out a safe habitat on the seabed among rocks and algae and spend the next 2 to 3 years there, regularly shedding their old shells. Roughly 4 to 6 years later, they become sexually mature and the whole cycle can begin again. Lobsters feed on invertebrates such as snails, mussels, starfish, and sea urchins, other small crustaceans, and worms. As larvae, they, in turn, are food for fishes. As adults, they land on plates in restaurants where, after having survived the ordeal of transport, they die a horrific death by being dropped alive in salted boiling water. In 2007, it was shown that lobsters are capable of learning and that they can feel pain, fear, and stress. In 2017, the Federal Administrative Court in Berlin recognized the capacity for suffering of lobsters and other crustaceans. When you consider the animal welfare regulations that exist to protect the farming, transport, and slaughter of mammals, every seafood fan should perhaps also think about the suffering of these creatures and do without lobster in the future.

Migratory Slitherers

There are fish that migrate thousands of miles just to mate for one single time before returning to their place of birth to die. Over the course of their lives, they change their appearance, shapes, names, and even habitat.

The migration of eels has fascinated biologists for ages. These animals, which are born at sea, spend most of their lives in rivers and lakes, only returning to the sea to spawn. Until now, 21 recognized eel species have been recorded exhibiting this form of migratory behavior. The life cycle of the eels is called catadromous (meaning migrating downstream to spawn), whereas many other fish, such as salmon, are anadromous (they spend most of their lives at sea only, migrating upstream to spawn). European eels (*Anguilla anguilla*) begin their migration in fall, between September and November. During their exhausting journey, they even cross stretches of dry land, slithering through moist grass, crossing the smallest of brooks and ponds, before eventually finding the relative luxury of a river to transport them toward the sea. Once there, they become active again by swimming. Their destination is the Sargasso Sea, a large sea current flowing clockwise in the Atlantic between Bermuda and the Azores. The ones that make it there have overcome many dangers—avoiding the turbines of hydroelectric dams, swimming through polluted waters, and escaping from the hungriest of all predators, human beings. Worldwide, eels are considered a delicacy, and because of their high fat content are well suited to smoking. In the meanwhile, the European eels are categorized as critically endangered. Up to now, all attempts at breeding and raising the animals domestically have proved unsuccessful. A number of research institutes have succeeded in rearing larvae, but none of the larvae have survived for more than 22 days. Reproduction and the early development of eels still remain great mysteries. The fact that we know anything at all about their fascinating life cycle is thanks to the Danish marine biologist Johannes Schmidt. At the beginning of the twentieth century,

Schmidt set out to find the spawning grounds of the eels, to try to shed some light on the puzzle, and met with success. Some 3,000 miles from the west coast of Europe, he identified the Sargasso Sea as the most likely spawning place of European eels. Here, he managed to fish freshly hatched flat, transparent eel larvae, leptocephalus larvae, out of the water. Directly after hatching, the small larvae head for the coastal waters of Europe. Once there, they go through a transition, eventually becoming roughly 3-inch-long glass eels. At this stage, they are also transparent but look a lot more eel-like. In spring, most of the glass eels then swim up the rivers of Europe. At this stage they are called elvers, and then later yellow eels (due to the yellow pigmentation of their bellies). Once they have chosen their habitat, they remain there until sexually mature—males in 6 to 9 years, and females in 12 to 15 years. Generally, maturity of European eels varies widely, and can be anywhere from 4 to 20 years, depending on the respective habitats. In the south, the tendency is toward earlier sexual maturity. The farther north they are, the longer it takes. Another anatomical adaptation to their diets splits the European eels into two varieties, each reflecting the food supplies of the respective habitats. Eels living mainly on small, soft-bodied prey such as small crustaceans, worms, and insect larvae tend to have narrower heads, while eels preying on larger, hard-bodied organisms such as fish, shrimps, and frogs have broader heads.

When the time comes to reproduce and migrate back to their place of birth, the eels' shape and appearance changes once more. The previously brownish-green color of their bodies transforms to a silvery gray; the eyes become larger; and the head is more pointed. Within a month, the whole digestive tract decreases in size to make space for the growing sexual organ. The eels, now called silver eels, eat nothing the whole way back to the Sargasso Sea. They use reserves of fat that they have built up from eating shellfish and other fishes over the course of their lives. Once the eels reach the Sargasso Sea, we lose all trace of them. The act of mating has never been observed in the wild. It is speculated that the first sexual act of the animal is its last. So, the final secret of the eels is, and remains—at least for the time being—a mystery.

ENDANGERED
BLUE WONDER

~~~~~~~~~

THE SEA SEEMS wild, beautiful, untamable, and infinite. Sadly, that first glance is misleading. The oceans are in great danger. From the tropical coral reefs to the polar seas, from the epipelagic to deep-sea trenches, from plankton to whales— marine ecosystems that have worked so well together for thousands of years have, in recent decades, become threatened. The reason is us—human beings. For hundreds of years, the sea was considered a boundless resource: supplies of fish, shellfish, and oil seemed limitless. We lived off the fat of the sea. We took and are still taking more than we need, more than necessary. But, hey, never mind. Nature can always restock, and who cares about the damage and garbage left behind anyway, right? Every year a day is dedicated to this greed: Earth Overshoot Day. Overshoot Day is the brainchild of the Global Footprint Network. It is intended to draw attention to the day in the year when the demand for regenerative resources exceeds the supply and capacity of the Earth's resources. In 1987, Earth Overshoot Day was December 19. By 2018, this had

shot forward to August 1. This means that in 2018, from an ecological viewpoint, we were living way beyond our means and
had consumed 1.7 Earths or, to put it differently, we had used
up nature's resources 1.7 times quicker than the ecosystem
needs to restore them. The situation in 2019 was very similar:
the key date was July 29.

We are extracting more than can be replaced, be it trees,
terrestrial animals, resources, or marine dwellers. Simultaneously, we are, ever more quickly, pumping increasing amounts
of CO_2 into the atmosphere, to the point where the forests
and seas cannot keep pace with absorption, causing wide-
ranging consequences not only for plants and animals but also
for us. There is even a term for our epoch—the "Anthropocene," derived from the ancient Greek word *anthropos*, meaning
"human." Our existence on Earth has had an influence on the
planet—and not a good one. The sea, an apparently inexhaustible ecosystem, is threatened from all sides: from the pollution
caused by oil and plastic, from the effects of global warming,
from the extinction of species through overfishing, and from
the destruction of marine habitats. It is high time to name the
hazards so that, if possible, we can avert further damage.

The Curse of Black Gold

It powers cars and other machinery, warms us and gives us
light...and it pollutes the oceans. Although it has been known
for thousands of years and was used by ancient civilizations,
wide-scale commercial extraction only really started in the
second half of the nineteenth century. It is one of the most
important resources for our industrial society and was formed

many millions of years ago. Of course, we are talking about petroleum. Before petroleum became the black mass it is now, it used to drift around the primeval oceans as tiny planktonic organisms, consisting mostly of algae. After dying and sinking to the ocean floor and then decomposing, deposits of putrid sludge were formed that, over the course of time, were covered by other sediments. Then, after various chemical and physical processes such as increased temperature and pressure, the black gold that today is extracted worldwide on land and sea was created. Not always without an impact on nature. One of the worst catastrophes occurred on April 20, 2010, in the Gulf of Mexico, when the offshore drilling rig Deepwater Horizon exploded, killing eleven crewmen. The explosion on the rig, leased by British Petroleum (BP), caused a huge fire and the most disastrous oil spill of all time. Despite all attempts to extinguish the fire on the high seas, the metal construction collapsed after 36 hours of heat exposure, and on April 22 the platform sank beneath the waves. After the platform disappeared, it was noticed that oil was gushing out from a number of places on the seafloor. It was 3 months before the leaks could be plugged, but by then roughly 4.9 million barrels of crude oil had spilled at a depth of 5,000 feet. An oil slick about the size of Cape Breton Island polluted a 50-mile stretch of the Louisiana coastline, meting out a death sentence to thousands of birds, marine mammals, fish, sea turtles, corals, and other sea creatures. Some of the oil gushing out of the leaks also dropped through the water column to the seafloor, polluting an area of some 7,000 square miles. Settling on top of the sediment, it buried all forms of life. These organisms were then unable to access food or oxygen due to the coating of thick tar, making microbial degradation of the oil impossible.

BP decided to disperse the oil spill on the surface by using a controversial chemical hammer to prevent the oil slick reaching the coast. Over 260,000 gallons of Corexit 9500 was used to break up the gooey oil mass into small droplets, which then sank to the depths of the Gulf of Mexico. Additionally, they tried scooping oil from the water's surface and scraping the film of tar off the beaches. Scientist and environmentalists, however, believe that a large proportion of the leaked heavy crude oil is still on the seafloor. Even now, particularly after storms, lumps of tar can be found washed up on the beaches of the five affected states. Animals are still suffering from the aftermath of the catastrophe. In 2014, four times the number of dead dolphins were washed ashore than the year preceding the disaster in the affected area. The animals had died as a result of lung diseases, kidney problems, and loss of teeth—all symptoms that occur on contact with oil. Volunteers involved in cleaning up also fell ill from the effects of using Corexit 9500. And as if all this wasn't bad enough, 22,000 people lost their jobs in the fishing industry because the fish were either dead or inedible, and oyster farmers went bankrupt.

The explosion on Deepwater Horizon, catastrophic though it was, was only the tip of the iceberg. Lower-profile incidents happen to supertankers at regular intervals, almost as if it has become routine. It is estimated that spillages of 100,000 tons of oil are leaked into the environment annually.

But crude oil isn't the only substance marring the oceans. In recent decades, another material made from oil—something so ubiquitous that it is difficult to imagine life without out it—has infiltrated the seas: plastic.

The Age of Plastic

We come across it everywhere: in the mornings when brushing our teeth, in the shower when using shampoo, later when checking emails on our smartphones while clutching refillable thermoses of coffee, and even when we climb into our cars, take a train, or jump on our bikes to get to work. In our technological world, it has become indispensable, and progress without it seems unimaginable. Plastic is everywhere. In many ways, it is a great invention. Depending on the requirements, it can be hard as steel but much lighter, transparent as glass but not as fragile, or elastic but not too easy to snap. Plastic has many unique properties and wide-ranging applications. It is cheap to produce; it is stable in regard to temperatures, light, and chemicals; and when mixed with other substances its hardness, strength, or elasticity can be regulated. Oil-based plastic started its conquest of the world in the early fifties with the beginnings of industrial production. Since then, the worldwide production of plastic has grown continuously, and by 2018 annual production had exceeded 400 million tons. Half of this is produced in Asian countries—above all in China—followed by Europe (including Norway and Switzerland) with 60 million tons and an upward trend. In Europe, plastic is mostly used for the packaging industry, followed by building materials, and then the automobile industry. Of course, with production increasing the amount of waste products are also on the rise. According to a study published by the World Bank in September 2018, 242 million tons of plastic waste was produced worldwide in 2016, which accounts for roughly 12 percent of all municipal waste. Naturally, it is not only plastic waste that is increasing but also the amounts of waste in general. Annually, 2 billion tons of waste are produced world-wide, and if you were to pile this unimaginable load onto trucks they would stretch around the world 24 times. If this tendency continues, we will have to reckon with an increase of 70 percent by 2050. In plain language, in just over 30 years waste production worldwide will swell to 3.4 billion tons.

Amounts of inorganic waste, like plastics, increase and organic waste decrease with rising incomes. According to the German Environment Agency (Umweltbundesamt, the UBA), in Germany alone, 18.2 million tons of packaging waste, both industrial and domestic, was produced in 2016—that is almost 500 pounds per capita. This means that Germany is way above the European average of 360 pounds per capita. Bulgaria, with a per capita waste of 120 pounds, produce the least packaging waste in Europe.

While many countries collect and either dump or recycle garbage, the infrastructure is missing in less developed countries. Even with the appropriate infrastructure, over 90 percent of garbage in developing countries isn't properly disposed of and pollutes the environment with far-ranging consequences. Within Europe there are serious differences in the treatment of plastic waste. In Germany, Holland, Belgium, Switzerland, Austria, and the Scandinavian countries, the proportion of plastic waste in the landfill sites is set at 10 percent, if not totally forbidden. The proportion in southern European countries rises to 50 percent. In Germany, for instance, waste is collected and recycled or used to create energy. According to the UBA, 6.15 million tons of total plastic waste (waste from industry and home waste) was produced in 2017; 2.87 million tons (46.7 percent) of the total waste was then recycled. But recycling there includes both the waste that is to be recycled in

Germany and used in the manufacture of new plastic products as well as waste that is to be exported to other countries. Almost 53 percent, 3.24 million tons, is burned instead of fossil fuels and used to generate power. The remaining 0.6 percent is stored in depots and includes, among other substances, construction wastes. Even though a large proportion remains in Germany, export of waste to foreign countries is a problem. In 2016 alone, Germany exported around 11 percent of all its packaging waste (including metal, paper, and plastic waste) to foreign countries, with more than half going to China, the world's garbage dump. Since the beginning of 2018, however, China has banned the import of unsorted plastic waste because of massive environmental intrusions. Germany, fearful of ending up with its own waste, is searching for alternatives in other countries like Malaysia, Indonesia, Thailand, and Vietnam. When I was in Vietnam in August 2018, I was able to speak to local authority agents and to university lecturers about waste management. They told me about the container ships in Vietnamese harbors laden with plastic, paper, and metal waste from Europe, Japan, and the United States waiting to be processed. As Vietnam didn't have the capacity to cope with the quantity over the previous 4 months, 28,000 containers had piled up in the country's harbors. In the Tan Cang–Cat Lai terminal in the harbor of the capital, Ho Chi Minh City (one of the largest ship terminals in the country), 8,000 standard containers (TEUS) full of plastic, paper, and metal waste were awaiting processing. Unfortunately, Vietnam already has grave waste problems and is hopelessly overburdened with the appropriate disposal of its own waste. Most of its disposal facilities don't fulfill the technical or ecological criteria. Of course, this raises questions about what happens to our waste when it gets there. As in China, transparency is lacking and the possibility of tracking imported garbage is tricky. Vietnam holds fourth place on the list of countries with the most plastic garbage reaching the ocean, behind China (1), Indonesia (2), and the Philippines (3). The Mekong is one of the ten rivers transporting the most plastic waste to the sea (see plate 23, bottom). It may be easy to point a finger at these countries, but as I have said, it is precisely there that the yogurt container you have just recycled is most likely to land. According to the U.S. Census Bureau, 78 percent of plastic waste in the United States (0.83 million metric tons in 157,000 large TEUS) was exported to countries with poor waste management in 2018.

If the yogurt container ends up somewhere in the middle of nature, then both the plastic and the additives the plastic contains (flame retardants, softening and stabilizing agents, and food fillers) become damaging. These are harmful to animals and the plant world, and eventually find their way into human beings. A prominent example is the case of bisphenol A (BPA), which acts a bit like the hormone estrogen in human bodies. BPA is one of the most-produced industry chemicals worldwide and can be found in all sorts of everyday articles like children's toys, CD/DVD blanks, thermal paper for receipts, polycarbonate bottles, cosmetics, cans, pacifiers, food packaging, and plastic knives and forks. BPA makes plastics harder and thus more durable. Even though representatives of the plastic industry claim the intake amounts of BPA in adults and children are tiny and insignificant, specialist doctors and scientists have long been sounding the alarm. Studies have found the harmful substance in popular canned foods such as tuna, tomatoes, coconut milk, and sweet corn. In March 2018, the European Union's chemical restriction legislation, REACH, placed BPA in the "very high concern" category. The World Health Organization also classified BPA as an endocrine disruptor, which means that it is a substance that can interfere with the body's hormonal (endocrine) systems. In animal experiments, it has been linked to the disruption of sexual and brain development, diabetes, sterility, and cardiovascular diseases, among other disorders. When BPA is leaked into the open environment, it could lead to the feminization of male fish, thus endangering the continuance of fish populations. The consequences for humans could mean developmental disorders, particularly in children; a reduction in sperm count; and even infertility. A higher risk of breast cancer for women is also being examined. People who are in regular contact with products containing BPA have higher concentrations of BPA in their blood. One example is cashiers in supermarkets, who umpteen times every day have to handle thermal paper receipts. Of course, BPA can reach the bloodstream through skin contact, too. Luckily, BPA-free thermal paper alternatives exist—they just need to be used. In Japan, foods have been sold in BPA-free cans for two decades now, and France has banned the use of BPA in all materials that come into contact with foodstuffs. Because babies and toddlers are especially vulnerable to harmful substances, the use of BPA in baby bottles has been banned by the EU and Canada due to health concerns. The U.S. Food and Drug Administration (FDA) ended its authorization for the use of BPA in baby bottles and infant formula packaging. According to the FDA, based on its most recent safety assessment, "BPA is safe at the current levels occurring in foods."

The harm begins once plastic, with its additives, reaches the open environment. Plastic, over time, becomes brittle under environmental influences and breaks down into ever-smaller pieces until it decomposes completely. This can take a while: a plastic bottle needs 350 to 400 years to completely break down. In the process, the plastic particles are not only releasing toxins but also accumulating them. Plastics, like a sponge absorbing liquids, bond environmental toxins and transport them from one ecosystem to another. This could be, for example, insecticides like DDT or the now-banned carcinogenic PCBS (polychlorinated biphenyls), which were widespread until the 1980s and are still present in the atmosphere, the soil, and waters worldwide. Animals that mistake this spiked garbage for food are about to absorb real chemical bombs. The toxins bonded to the plastic are released inside the bodies of the animal and become concentrated in the tissues. If the animal then ends up on the plate in front of us, then we too will then consume the highly toxic contaminants. And there is another problem: once the most often used plastics—polyethylene and polypropylene—are released into the open environment, their decomposition produces greenhouse gases, thus accelerating global warming.

Are Bioplastics the Solution?

So, what are the alternatives to conventional plastics? Is bioplastic the solution, and what exactly is it? The term "bioplastic" is already misleading, and since the name hasn't been patented, it isn't used in a consistent manner. At the moment, bioplastic includes all biodegradable polymers as well as ones that are not necessarily only produced from renewable resources. But bioplastics also include polymers based on renewable resources, which in turn are not necessarily biodegradable. It's complicated and confusing!

In order to be classified as biodegradable and be labeled compostable, products in Europe have to conform to norm EN 13432. This means, among other things, that after 3 months in a compost heap no more than 10 percent waste materials can be left after sifting with a 0.8-inch mesh sieve. Also, chemical residues have to be within preordained limits and not have harmful effects on fauna and flora. Biodegradability in water-based media requires a 90 percent transformation of organic materials in CO_2 within 6 months.

In the United States, the regulatory framework is the ASTM D6400 standard, which has similar requirements to the European norm EN 13432. The certification is carried out by the Biodegradable Products Institute (BPI). The BPI is "North America's leading certifier of compostable products and packaging." If you wish to learn more about products that are truly compostable, visit their website at bpiworld.org.

When looking for compostable garbage bags at your local supermarket, you will likely find some claiming to be biodegradable. On closer inspection of the fine print, you will often discover that they contain oxo-degradable and photosensitive plastics like polyethylene. These bags don't comply with the above-mentioned standards. They contain additives that simply allow them to decompose to small particles. In addition to the plastic fragments that can disperse in all possible directions, the additives also contain toxic substances that are then released. If, on the other hand, you are looking for bioplastic compost bags and dispose of them in the compost containers, they have to be carefully sorted by hand at the waste treatment centers into different categories: 1) bags that degrade slower than the organic waste, and then only under certain physical conditions via fungi or microorganisms; 2) bags that have no value whatsoever as compost; and 3) bags that are difficult to distinguish from non-compostable bags and that then have to be added to the non-recyclable waste, which at least means they will be of some use as energy. Products made of bioplastic that reach the open environment create as much of a problem as "normal" plastic. Bioplastics require special conditions, such as a certain temperature, humidity, and oxygen, plus micro-organisms, to decompose. However, it may take months for these conditions to accrue, and it really doesn't matter to the animals dying in the rest of the garbage whether the straw, bottle top, or yogurt cup is made of bioplastic or ordinary plastic—the result, sadly, is the same.

Polluted Paradises

A riverbank, a roadside, a local beach, or a tropical paradise far, far away, all have something in common—plastic garbage. My work has allowed me to visit a number of wonderful places on Earth, and for the most part I keep my head under water. I often think back fondly to my time in the Maldives, snorkeling in the flat inner reefs watching octopuses mate (see plate 20, bottom) or swimming on the reef crest, face-to-face with a group of inquisitive blacktip reef sharks, or watching a hawksbill sea turtle taking a nap beneath table coral (see plate 6, top)—moments I will never forget. Too often, however, objects that shouldn't have been there intruded upon my underwater panorama. I don't need to explain how sad it is to see a coral suffocate after becoming engulfed by some diapers, a plastic bag, or an old fishing line. Honestly, sometimes parts of the underwater world of the Maldives looked like a garbage dump. I was constantly gathering trash and bringing it ashore.

Of course, this is not the image that you will glean from glossy magazines at travel agencies, and it can be very shocking when you come across it for the first time. Paradise is polluted. The Maldives is lacking an efficient waste management system. A considerable amount of garbage that the tourists bring is either burned or ditched at sea.

But the improper disposal of plastics isn't restricted to the Maldives. It's a worldwide crisis. Garbage is thrown overboard, blown by winds to the sea, or transported downstream by rivers. Globally, between 4.8 and 12.7 million tons of plastic waste land in the sea every year, which, on average, is the equivalent of one garbage truck load of plastic *per minute*. This enormous amount is estimated to double by 2030, and even that sum is estimated to quadruple by 2050 unless influx is drastically reduced. Simply put, in 2025 there will be 250 million tons of plastic garbage in the sea or 1 ton of plastic waste for every 3 tons of fish. And if trends continue, by 2050 there will be more plastic swimming in our oceans than fish. The greatest influx of plastic—95 percent—reaches the sea via rivers. Ten rivers are the main culprits, together washing 4 million tons of plastic downstream to the sea every year. Eight of these rivers are in Asia, with the Yangtze topping the list; two are in Africa. The remaining 5 percent of plastic waste comes mostly from the illegal dumping or loss of freight by ships.

All this waste then spreads throughout the oceans and throughout the water column. Japanese researchers found a plastic bag in the Mariana Trench at a depth of 35,700 feet and more than 600 miles from the nearest land. A total of

3,425 pieces of plastic waste have been registered in the data records and photos from over 5,000 ROV dives over the last 30 years, of which 89 percent were disposable products. Potato chips packaging was found in the stomach of a coelacanth. Coelacanths are living fossils—of which there are only two existing species—that live at depths of between 300 and 1,200 feet. They have been swimming through the oceans almost physically unchanged for 400 million years, and now they are suffering from garbage that we have been producing for only a couple of decades.

Even with most of the waste sinking to the depths of the ocean, at least 5.25 trillion plastic fragments with a total mass of 269 million tons are floating on the surface. They are transported hundreds of miles by the currents, accumulating in oceanic garbage patches (more on that soon). A large proportion of the plastic garbage on the surface and in the water column consists of ghost nets. Ghost nets are fishery nets that have either been illegally dumped in the sea or lost during rough weather. According to the United Nations, every year some 640,000 tons of nets are lost at sea, which doesn't seem to force fishermen to give up fishing but simply to start all over again. These ghost nets then become floating death traps for countless sea birds, marine mammals, sea turtles, sharks, and other fish, causing slow and painful deaths. Some of them—the size of soccer fields—become snagged on reefs, suffocating the corals beneath them. Removal of such nets from reefs proves difficult, as other corals grow through the mesh and break off during the removal process.

When I was working in the Maldives, I was often called upon to free animals trapped in ghost nets, especially during the northeast monsoons. The nets were not from local fishermen (this form of fishing was banned there) but had drifted into our waters from Southeast Asian countries such as India, Sri Lanka, and Thailand. The northeast monsoon, which Maldivians call *Iruvai*, marks the dry season on the tropical islands in the Indian Ocean. The monsoon usually lasts from November to April, and with the sunny weather and calm seas attracts the most tourists. But with the good weather come the nets, which mostly trap sea turtles. Of the seven species of sea turtles, which are all threatened with extinction, five live in the waters around the Maldives. The creatures most often caught up in the ghost nets are olive ridley sea turtles (*Lepidochelys olivacea*), one of the smallest species of sea turtle. I remember well the first two animals I freed from nets. One afternoon, I was alerted by the staff of the hotel to which the marine station also belonged: a net had washed up on the beach with two helpless olive ridley sea turtles entangled in it. Very carefully, we cut the animals free of the heavy 15-foot-long net. Exhausted, the frightened creatures put up very little resistance. After we had transported them to the pools in the marine station, we took a closer look at their wounds. The smaller turtle had only superficial grazes, while the larger one was much more gravely injured. The net had dug so deep into her flesh that the front flipper had become seriously inflamed. There was also severe chafing of the neck, probably from desperate attempts to escape the netting. After disinfecting the grazes, we released the smaller turtle and watched her dive into the blue of the Indian Ocean. We planned to take the more seriously injured turtle to the Olive Ridley Project (a marine rescue center based on another island) the next day. Sadly, the turtle didn't survive the night. So that the death wasn't totally pointless, we buried the creature in the sand

and, a couple of months later, dug it up in order to rebuild the skeleton and exhibit it at our Eco Centre (see plate 21). A different turtle rescue luckily had a happier ending. Weeks later our diving boat found another net in the water with a trapped turtle in it. The net had dug into the front left flipper so deeply that it was only connected to the rest of the body by a small piece of tissue. The flipper was also completely swollen and the wound margins severely inflamed. As I'm not a veterinarian and didn't think that I could perform an amputation, our regular doctor gave it a tranquilizer, and we treated the wound as best we could. Early the next morning, we went by speedboat to Malé, the capital of the Maldives, and from there took another boat to the turtle hospital, where the marine biologist performed the amputation of the front flipper. On seeing a fountain of blood jet out of the wound, I thought that the animal wouldn't survive the operation. But by some miracle, the turtle was released to the ocean two weeks later, with one less flipper but alive.

It is not only the large nets posing a threat to sea turtles but also the pieces of plastic that the animals mistakenly eat. The bits of plastic are like ticking time bombs for sea turtles. Chris Wilcox and his team calculated that the risk of death after swallowing one single plastic fragment was 1:5. If the number of plastic fragments swallowed rises to fourteen, then the chances of death are 50:50. If two hundred particles are swallowed, certain death follows. The authors of the study estimated that roughly half of all sea turtles in the world have ingested plastic—the figure even rises to 90 percent for juvenile green sea turtles off the coast of Brazil. The anatomy of the sea turtles' digestive system doesn't allow them to regurgitate plastic and disgorge it. Plastic fragments can block the digestive tracts and cause internal wounding, which can prove lethal.

In addition to ghost nets and all the other plastic fragments, plastic bags are perilous for sea turtles, as the floating garbage can easily be mistaken for the sea turtles' main food source: jellyfish. Confusing plastic bags with jellyfish is also often the downfall of the bizarre-looking ocean sunfish (*Mola mola*), if not through suffocation then from starvation when the plastic bags clog the stomach.

At the end of 2018, a dead sperm whale was washed ashore on the Indonesian coast. A post-mortem examination found that the animal had 13 pounds of plastic in its stomach: 25 plastic bags, 115 plastic beakers, 2 flip-flops, a nylon sack, and 1,000 other plastic fragments.

Plastic Patches

Debris that lands in the sea is transported by the sea currents, the vast majority of which sinks to the ocean's depths. The remaining rubbish is either washed up on the beaches or floats on the surface, eventually accumulating in high concentrations in the open sea. The media came up with the term "plastic slick," but this suggests a continuous expanse of plastic waste. These garbage slicks are more like plastic confetti or plastic smog just below the water surface; larger fragments are seldom represented. The largest of these accumulations of plastic waste is located between California and Hawaii. Known as the Great Pacific Garbage Patch, it holds an estimated 79,000 tons (at least) of plastic. This gigantic amount of debris is spread over an area roughly three times the size of France.

Almost half of the debris (47 percent) consists of derelict nets, ropes, and other miscellaneous bits of fishing gear. The rest of the 1.8 trillion plastic fragments are mostly microplastics. Large slicks of plastic waste and ghost nets also gather in circulating currents called oceanic gyres. There are five of them: in the Indian Ocean, the North Atlantic, the North Pacific, the South Atlantic, and the South Pacific. Oceanographers also refer to them as the "Big Five," and these gyres help power the ocean's conveyor belt, which circulates ocean waters around the world. Debris that accumulates in the Big Five eventually form slicks.

Microplastics—
The (Almost) Invisible Danger

The longer plastic floats in water, the more brittle it becomes. The more brittle plastic is, the easier it becomes to break down from the effects of friction, the motion of waves, biological processes, and UV light into increasingly small pieces of microplastic, which scientists define as anything smaller than 0.2 inches. Another big problem is microbeads—small granules of plastic found in cosmetic and personal care products, in drugs, in the paints used on boats and for road markings, in the materials used in sandblasting, in the plastic fibers of synthetic clothing. All of it ends up in wastewater. Unfortunately, these microbeads, usually invisible to the naked eye, cannot be completely filtered out at wastewater treatment plants, and so are flushed into our rivers and, eventually, the sea. Once in the sea, microplastics are distributed (in)opportunely, mixing with living plankton. Over 90 percent of all the plastic reaching the ocean eventually sinks to the depths, never to be seen again. Down there on the seafloor, concentrations of plastic are many times greater than those on or near the surface, and particles settle there as new sediment. In sediments in the Mariana Trench, the deepest known point on our planet, between 750 and 8,300 microplastic particles per gallon have been recorded. The only other deep-sea area with higher microplastic abundances were found in the Hausgarten deep-sea observatory in the Arctic Greenland Sea west of Spitsbergen. There, a research group from the Chinese Academy of Sciences recorded levels of 12,850 microplastic particles per gallon. If that is not a broad hint, I don't know what is.

It is really difficult to imagine the sheer amounts of plastic waste, big and tiny, in our seas. For this reason, a group of environmental activists (including myself) and windsurfers set up the Aquapower Expedition in 2015. The idea came from the German windsurfing pro and activist Florian Jung. During the expedition—a mix of water sports and science—we planned to make the slick of debris in the North Atlantic, which at first glance seems invisible, visible. With the aid of web episodes and daily updates from on the water, our goal was to convey the facts about the sea as a habitat as well as the threats to it in an easily understandable way. In many ways this trip was one of the most beautiful, exhausting, and fascinating things that I have ever experienced. It showed me once more how important the sea is for us and how little we care about the basis of our existence. In a 50-foot-long catamaran full of surfing, diving, and scientific gear, we sailed, in 72 days, 5,290 nautical miles from Guadeloupe in the Caribbean to the Virgin Islands, and from there to the Dominican Republic and Bermuda through the North Atlantic garbage patch to the Azores, docking finally in Marseille on June 8, World Oceans Day.

During the expedition, we regularly cast our plankton net (see plate 22, top), and no matter where we took samples—the Caribbean, the Atlantic, or the Mediterranean—every single sample was full of microplastics: fragments from larger plastic articles, polystyrene, scraps of plastic foil, synthetic fibers mixing with plankton to make a soup (see plates 22, bottom, and 23, top). Organisms that feed off plankton are not able to distinguish and sift through the soup for their nutrition, so they swallow the lot.

In 2015, a study carried out by Chelsea Rochman's research group compared edible fish from the fish markets of Indonesia and the United States, and came up with an interesting result: there were mainly plastic fragments in the stomachs of the Indonesian fish, whereas the U.S. fish and mussels contained mostly plastic fibers. The study group believed that the difference in stomach content reflected the different waste management strategies. While in Indonesia and Vietnam most of the plastic waste end up in the sea and breaks down there, most of the waste in the United States is recycled, dumped, or burned. The microplastic fibers in the stomachs of fish and mussels from the U.S. markets probably originated from washed clothing, as 2,000 microplastic fibers per wash cycle end up in wastewater. When we remember that microplastics are bearers of stockpiled toxins and pathogens that can be absorbed by animals and accumulate in tissues, then the fish on the plate in front of us only tastes half as good. It is speculated that people who eat seafood regularly ingest up to 11,000 microplastic particles per year. Even though humans excrete most of the microplastics over the course of time, the possible ill effects on health cannot be excluded with any certainty.

During the expedition, I had a particularly alarming experience off Bermuda on the way to observe humpback whales with Choy Aming, a local filmmaker and shark and whale expert. Spring attracts thousands of humpbacks from the Caribbean, where they give birth to their calves, to Bermudan waters in the North Atlantic, where they come to feed. Humpbacks feed mostly on plankton such as krill but also eat small fish, filtering them from the water with their baleen plates. On that day we were in a small boat looking for humpbacks crisscrossing the Bermudan waters when we came across an area full of brown algae, genus *Sargassum*. On closer study, we noticed that the algae cluster was flecked with color. Curious to know the cause, we took a bucketful as a sample. As I withdrew my hands from the water with the bucketful of algae, they too were covered in colored particles. It was plastic. And it is precisely these plastic particles that the humpbacks must ingest when feeding on plankton. Assuming these particles release toxins in the animals, then it is not difficult to imagine that the suckling calves also ingest toxins via the mother's milk. The fourth episode of the wonderful BBC series *Blue Planet II*, broadcast in 2018, featured a grieving long-finned pilot whale cow carrying her dead calf in her mouth for days. The broadcaster and natural historian Sir David Attenborough speculated that the newborn calf had probably been poisoned after suckling contaminated milk and that it was likely these toxins had originated from plastic particles or other environmental toxins. This was only speculation and hadn't been tested or verified, but on the basis of other scientific evidence it seems to me to be highly plausible.

The Marine Megafauna Foundation studied the effects of microplastics on large filtering animals such as manta rays, whale sharks, and baleen whales, which swallow probably hundreds of (micro)plastic particles every day. Their conclusion was that the filtering of indigestible plastic particles can block the absorption of nutrients and damage the digestive tracts of the animals. Moreover, the chemicals associated with plastics and toxins can, over decades, accumulate in the tissues affecting biological processes such as growth, development, and reproduction. The main author of the study, Elitza Germanov, is certain that with the intake of contaminated microplastics, the populations—particularly of long-living species—will continue to decrease, as they will be producing fewer offspring. She also added that it is important to understand the impact of microplastics on the health of the giants of the sea, as almost half of all manta rays and devilfishes, two-thirds of filter-feeding sharks, and more than a quarter of baleen whales are classified by the International Union for Conservation of Nature as globally threatened species, making their protection a priority.

The floating plastic debris is not only harmful to sharks, whales, and other marine organisms, it also kills more than a million seabirds every year from clogged stomachs. Some calculations even predict that if trends continue, by 2050 as much as 99 percent of all seabirds will have ingested some form of plastic. This is hardly surprising—in some parts of the oceans there are already more microplastic particles floating around than plankton.

Climate Change and the Sea

The ever-increasing amounts of waste from our growing consumption contributes to the littering of our oceans as well as the global increases of greenhouse gases in the atmosphere.

Even though not everybody seems to have realized it, climate change is not fiction, and the effects are already being felt. The concentrations of greenhouse gases in the atmosphere, according to measurements from climate research institutes, are higher than ever: Records from the World Meteorological Organization (WMO) in Geneva show that the average global CO_2 concentrations in 2017 of 405.5 parts per million (ppm) are 146 percent higher than pre-industrial (pre-1750) levels (280 ppm). These were the highest CO_2 levels in the last 400,000 years, for which we have good data. Even though *Homo sapiens* only evolved some 200,000 years ago, climatologists can draw these conclusions from ice cores taken at depths of almost 2 miles from ice sheets in Greenland and Antarctica. Within these ice core samples, age-old trapped bubbles enable us to go back in time and see what the atmosphere and climate were like in long-gone eras and how they have changed over time (akin to dendrochronologists reading the growth rings of certain trees). During the Iron Age, CO_2 levels were roughly 200 ppm, while during the interglacial periods they were about 220 ppm. In 2015, for the first time in history, levels topped 400 ppm. In May 2018, they were just above 412 ppm. The relentless rise of carbon dioxide levels in recent times is notably in a constant relationship with the amount of fossil fuels burned and has catastrophic consequences for human health and the planet. Should the present trend continue, the average global temperatures will, in the long run, rise 7° to 9°F. Through the massive consumption of fossil fuels for industry, heat and electricity production, transport, and agriculture, more CO_2 will be pumped into the atmosphere than nature can absorb. Carbon dioxide in the air absorbs longwave thermal radiation that is usually transferred to space. If there is

202

too much of the gas in the atmosphere, this will lead to gradual global warming. Carbon dioxide is responsible for roughly three-quarters of this human-made greenhouse effect. "Without rapid cuts in CO_2 and other greenhouse gases, climate change will have increasingly destructive and irreversible impacts on life on Earth. The window of opportunity for action is almost closed," stated Petteri Taalas, secretary-general of the WMO. Research shows that increased CO_2 levels, if not regulated, can cause tens of thousands of deaths, increases the number of heatwaves and superstorms, acidifies the seas, and causes a rise in sea levels.

The world community is aware of the risks and has met regularly since 1995. At the annual Conference of Parties (COP), a UN framework convention in Paris in 2015, the 195 member states, bound by international law, agreed to try to limit global warming to 3.6°F—if possible, to 2.7°F—relative to preindustrial levels. The national climate protection objectives are binding, with each country framing, communicating, and deciding on the measures. Every 5 years each country has to achieve new, ambitious goals to show headway in the limitation of global warming. Unfortunately, Germany failed to reach its stated climate protection target of reducing its greenhouse gases. According to the German government's climate protection report, CO_2 emissions will only be reduced by 32 percent (compared to 1990), instead of the stated target of 40 percent by 2020. This ranks Germany eighth in the climate protection ratings of the twenty-two COP member states in the European Union. "Despite a number of advances in energy transition in the past years, Germany's climate protection targets have generally not been achieved," stated the German Environment Minister Svenja Schulze, who promised: "We will learn from the mistakes of the past so that Germany will not fail to reach its targets again. We need more courage and commitment in climate politics, which is why I plan to submit a Climate Change Bill legally binding observance of our climate targets."

The outcome of a recent world climate convention, COP 24, in Katowice, Poland, in December 2018, was both a success and a disappointment. A success in that the world community finally ratified the implementation of the climate agreement presented in Paris. The ensuing policy formulated in Katowice is a kind of user's manual, specifying which methods are to be used for measuring CO_2 emissions and when the results have to be presented. The regulations are intended to increase transparency and strengthen the updating of the climate policies of the individual states until 2020. A disappointment, on the other hand, because of the unwillingness of COP 24 to "welcome" the latest key study of the Intergovernmental Panel on Climate Change (IPCC)—the conference decided to "take note" instead. The IPCC assessment report, compiled by ninety-one leading scientists from forty different countries, was drawn up at the request of the signatories of the Paris Agreement. The experts sifted through over 6,000 studies before presenting a summary. The findings presented a grim picture: limiting global warming to 2.7°F relative to pre-industrial levels is only attainable with immediate and forceful action. Measures for the drastic reduction in CO_2 emissions have to be introduced at once. By 2030, CO_2 emissions need to be reduced by 45 percent of the 2010 levels, and by 2050 down to zero.

We are already experiencing the effects of global warming 1.8°F above pre-industrial levels—think about the accumulation of extreme weather situations in recent years. According to the WMO, 20 of the hottest years since the beginning of records have been in the last 22 years, the top 4 all being in the last 4 years. The IPCC report stated that global warming between 2030 and 2052 will reach 2.7°F if temperatures continue to rise at the present rates. A global "Heat Age" will begin. These prospects should cause quite a bit of furrowing of brows, as we will then be experiencing a climate shift the likes of which we have never had before.

The Greenland and Antarctic sheets are also beginning to feel the heat, and the melting rate has increased fivefold in 5 years. If the Greenland ice sheet—at its thickest point almost 10,000 feet deep—melts, it alone could cause sea levels to rise by 23 feet in the next 200 to 300 years. And if that all seems to be in the distant future, there are countries already suffering from rising sea levels. Countries like the Maldives, which is only a few inches above sea level, have to seek out solutions to problems not of their making. In the last couple of years, populations from a number of Maldivian islands have had to be resettled, as their home islands have become uninhabitable. Strong storm surges have led to severe erosion of the beaches and the salination of freshwater reservoirs. The islands of the Maldives are not particularly large, and the people living there are linked very closely by family and social bonds. Resettlement has torn these communities apart, and there have been and are conflicts between the new settlers and the local populations on the islands. The sea, too, has to be reclaimed to provide land for housing the population. This means pouring sand onto the very coral reefs that protect the islands from the whims of the seas—basically, what the hotels have also been doing this whole time to the detriment of the coral reefs. But the impact of climate change has left its traces all around the world: torrential rainfall and flooding also cause homelessness, and heat and droughts have serious consequences for health and agriculture. Extreme weather is becoming more and more unpredictable and dangerous.

Even if we manage, in a united action, to remain within the 3.6°F limit, as stated in the Paris Agreement, it doesn't mean that we have overcome climate change. Things could even get worse: some areas of the world harbor "tipping points" in the climate system where irreversible consequences can be triggered and which could also lead to chain reactions. In the northern hemisphere, the melting of permafrost (the ground that remains frozen for at least 2 years) would result in the release of huge amounts of greenhouse gases. The permanently frozen ground preserves, rather like a deep freezer, plants, animals, and pathogens that have been iced in for thousands of years. If this ground melts, the organic materials decompose, releasing carbon dioxide and methane gas, which accelerate climate change. Other tipping points, such as the melting of ice at the poles, also fuel climate change and cause a rise in sea levels. Climbing temperatures lead to more fires and heat stress for trees in the other great carbon reservoirs, the forests. When forests burn, then the carbon dioxide stored in them is again released, heating up the atmosphere. This is not some distant dystopia. Glaciers and icebergs are now melting, the ground thawing, trees dying. If global atmospheric temperatures continue to rise, it will start a chain reaction that cannot be stopped.

Another factor in the inexorable rise in CO_2 levels comes from a totally unexpected quarter—plastic waste. In addition to the production of the products that later land on the disposal dumps, the waste itself creates greenhouse gases.

It has been discovered that the two most common plastics, polyethylene and polypropylene, when exposed to solar radiation, produce two greenhouse gases: methane and ethylene. Polyethylene is the most frequently produced and the most recycled synthetic polymer, as well as the most prolific emitter of the two gases. So, plastics are an (until recently unknown) source of climate-relevant trace gases that could increase with growing production and accumulation in the environment.

Under water, too, the impact of climate change is being felt very clearly. Marine calcite organisms, in particular, are suffering from the effects. The oceans, together with the forests, are the most important carbon sinks. Since the beginning of industrialization, oceans have absorbed about 30 percent of anthropogenic CO_2 emissions, the equivalent of roughly 524 billion tons of CO_2 or 22 million tons of CO_2 per day. The sea plays such an important role in worldwide climate: by absorbing CO_2, the sea slows down global warming. Absorbed CO_2 reacts with seawater, forming carbonic acid, and water becomes acidic, meaning its pH levels sink—the process is termed ocean acidification. Since the beginning of industrialization, the average pH ocean surface levels have decreased by 0.1 units, and now lie at 8.1. This equates to a rise in acidity values of some 30 percent. Increasing acidification reduces concentrations of carbonate ions, which are vital for the survival of calcareous organisms like mussels, snails, corals, and calcifying algae such as coccolithophores. These marine calcifiers need carbonate ions to form their shells and skeletons. When there are fewer of these molecules available, things get difficult for the organism, as the energy they have to burn to make up for the now more complex task of producing carbonate is unavailable for other processes. But even organisms that don't calcify (non-calcareous organisms) suffer from ocean acidification, as they have to use more precious energy to maintain their bodily functions. This causes problems in development, growth, and reproduction, as well as resistance to diseases and environmental stress. In a "business as usual" future scenario where CO_2 emissions remain at today's levels, the atmospheric CO_2 concentrations would be almost 930 ppm by 2100, compared to 400 ppm today. If things continue as they are today, pH values of the ocean will further sink 0.3 to 0.4 units by 2100. As pH scales are logarithmic, seawater will become 100 to 150 percent more acidic with a pH of 7.7. This doesn't mean, however, that seas will become a bubbling acid bath, dissolving all life within it. It simply means that seawater will be more acidic than it was in 1800, as a pH reading of 7.7 is still within the alkaline sector.

Ocean acidification, however, is not the only evil making life difficult for sea dwellers. Rising water temperatures and sinking oxygen levels are not making their lives any easier. These three factors alone are already difficult enough to overcome; in conjunction many marine organisms will have great difficulty in adapting to the new living conditions, and it remains to be seen which ones manage and which don't. Adaptation requires, above all, much time, and time is a commodity that is lacking.

The Death of Coral Reefs

Coral reefs are among the ecosystems on our planet with the richest variety of species. They suffer, in particular, from changes in conditions like ocean acidification and rising sea temperatures. In the seemingly idyllic sea's nurseries, it is especially easy to observe how a number of stress factors—not only climate change but also the fishing industry, diseases, tourism, and pollution—combine and to follow the far-reaching consequences of the death of corals on the whole marine ecosystem and, eventually, mankind.

Global warming leads to a rise in sea temperatures. For organisms like the coral polyps that need a certain optimum temperature to survive, the effect of the warming water is fatal. Corals bleach when suffering from heat stress. Coral bleaching leaves in its wake a graveyard of dead ghostly white skeletons of stony corals. But what exactly happens when corals bleach? If the ambient waters become too warm, the zooxanthellae (the corals' algal lodgers) are evicted from the polyps. When under heat stress, the algae produce toxins and the coral polyps, to protect themselves from these toxins, show their otherwise peaceful tenants the door. As the algae are the colorful parts of the partnership, the corals become white without their symbionts (see plate 6, bottom). Once a coral appears bleached, it doesn't necessarily mean that it is dead. Without their tenants, who provide up to 90 percent of their energy needs, coral polyps can survive a few days but are under increased stress and death rates rise. If the water fails to cool down for a couple of weeks, then the coral polyps will not accept new algal colonizers and starve. A 1.8°F increase in average summer temperature is enough to trigger mass coral bleaching. This form of coral bleaching has increased in strength and extent with climate change, and the intervals between bleaching incidents are becoming ever shorter. At the beginning of the eighties, bleaching occurred every 25 to 30 years. The gap since 2010 is down to a mere 6 years. Terry Hughes and his team from the James Cook University in Townsville, Australia, arrived at these figures in 2018. The time between bleaching events is too short to completely restock a reef community. The study concluded that fast-growing coral species need, as a rule, 10 to 15 years to recover, but the entire reef community will take much longer.

When a reef bleaches, then gradually the fish, sea turtles, crustaceans, and many other organisms based there disappear. But some species profit from the death of coral reefs. Algae, no longer bothered by pesky fish making a meal of them, can dominate the reefs. What remains is a slimy brown underwater desert with the occasional stray fish swimming by.

With the death of a coral reef, another climate-related risk factor comes into play. Ocean acidification also makes the life of the calcite-forming corals difficult. Sinking pH values mean fewer carbonate ions (the building material of their skeletons) are available, which can lead to the erosion of these skeletons. Additionally, acidic seawater clogs up particularly sensitive coral larvae, which then fail to develop healthily. As a result of this, fewer coral larvae become established and develop into colony-forming adults. Currently, one-third of all shallow water coral reefs are beyond reclaim, one-third are considered vulnerable, and just one-third are mostly intact.

The worst case of bleaching in the Maldives took place in 1998, with 90 percent of coral near the surface being affected. Water temperatures had been 1.8° to 5.4°F above average for a number of months. To this day, most of the reef hasn't completely recovered, and recurrent bleaching events make regeneration that much more difficult. Various coral species that can quickly settle and are fast growers can form a new reef in 10 to 15 years or resettle the old reef and thus provide the basis for the regeneration of a fully functioning reef community. In the case of the Maldives, it was mostly stony corals of the genus *Acropora*. Since 2014, however, clusters of bleaching events off the island state have affected 60 to 90 percent of the coral coating.

Sadly, 2014 proved to be a bad year for corals worldwide. The National Oceanic and Atmospheric Administration, the American scientific agency, spoke of worldwide coral bleaching lasting until the end of 2016. On the Great Barrier Reef, bleaching destroyed 30 percent of the corals on the 3,863 individual reefs that constitute the barrier reef during a 9-month ocean heatwave that occurred between March and November.

Coral bleaching events often occur in conjunction with El Niño, a climate phenomenon that causes above average water temperatures in many ocean regions every couple of years. The combination of El Niño and climate change means that coral reefs have less and less time to recuperate. The future looks pretty grim for tropical coral reefs with continuing climate change. If the world community fails to cap temperature rises at 3.6°F, then by 2100 the majority of the "rain forests of the sea" will have died. And when corals die, then so too do the countless other organisms living on the reefs. Furthermore, there is another man-made danger for the coral reefs—tourism. Sadly, not only do far too many people act like bulls in a china shop in the reefs, stomping over corals or breaking them off when diving or snorkeling, but they are unwittingly killing polyps with skincare ointments. Sunscreen is pure poison to sensitive polyps. Certain ingredients in sunscreen, such as oxybenzone, octocrylene, and paraben, are lethal to corals and other reef dwellers, even in minute quantities. These three substances in particular accumulate in small coral polyps and their larvae, and can cause development and growth disorders as well as bleaching. Annually, 4,000 to 6,000 tons of sun lotion reach the marine environment via water enthusiasts.

By now you likely won't be surprised to learn that plastic waste also presents a problem for corals. Joleah Lamb and her research team at the University of California studied 159 coral reefs in the Asia-Pacific region and found, believe it or not, 11.1 billion items of plastic entangled among the corals. And even this vast amount is supposed to increase by 40 percent in the coming 7 years! The scientists discovered that almost 90 percent of the 124,000 corals analyzed were more susceptible to diseases once they had had physical contact with plastic items. Plastic not only blocked out the sun rays vital for the survival of zooxanthellae, it also impeded the oxygen intake of the polyps. In another study published in 2017 in the *Marine Pollution Bulletin*, Austin Allen and his colleagues discovered that coral polyps actively catch the tiniest plastic particles and eat them. They even preferred them to their natural diet! Even though the polyps spat out the particles after 6 hours, there were still traces of some plastic particles in them. Further studies are needed to describe the precise effects of absorbed plastics on coral polyp. It is also important to discover which of the chemical substances were so irresistibly attractive to the polyps so that in future they can be avoided. Alarming findings, regardless!

Other threats range from nitrogen input from ever-intensive agricultural practices to an influx of industrial and domestic wastewaters. Corals are becoming more prone to diseases and less able to defend themselves from natural enemies or competitors for space. Also, especially in coastal regions, there is much construction happening (just think of the travel industry), and the sediment caused by such upheavals settles on fragile polyps, choking them.

In addition to their natural enemies, such as starfish, fish, and snails (see plate 15, bottom), blast fishing is an added burden for the reefs. In many Asian countries, the Caribbean and Africa, and sometimes the Mediterranean region, this destructive method of fishing is still widespread. Homemade explosives are thrown into the water, where they sink and explode, causing untold damage to reefs. The shockwaves do not discern between edible or ornamental fish, sea turtles, or other reef creatures—everything in the vicinity is literally blasted to bits. The edible fish that float to the surface are collected and sold on the markets; the rest are just collateral damage.

Last but not least, saltwater aquarium enthusiasts, knowingly or not, are also actively involved in coral bleaching and the disappearance of species in tropical reefs. As tropical fish, with a few exceptions, don't reproduce in aquariums, 99 percent of fish found there have been caught in the wild. Ornamental fish are cheap in countries such as Sri Lanka, the Maldives, Indonesia, the Philippines, Fiji, Australia, and Palau. The fish are stunned using highly poisonous cyanides like sodium cyanide and then collected: The poison is put into squirt bottles and taken under water, and then sprayed into small cracks in the reef. The stunned fish, now defenseless, are collected and brought ashore. There they vegetate in a bucket full of low-oxygen seawater before being tipped into a small plastic bag with lots of other fish and even less oxygen to be transported farther. Up to 80 percent of the fish don't survive the torturous trip from reef to specialist store. The remaining fish, now swimming in the aquariums of "enthusiasts," are by no means over the worst. The stress of different changes of water during an odyssey that often lasts weeks, as well as the cyanide residues, leave their marks, and the creatures may still die in the coming weeks or even months. And if the little fish dies, they can simply be replaced. In this way, some 500 to 600 million ornamental fish disappear from the reefs annually. People in the aquarium business in Germany can't complain about the lack of demand—80 million reef fish end up in German aquariums each year. Particularly after blockbuster films, like *Finding Nemo* or *Finding Dory*, there is a brisk trade in exotic fish. Even though clownfish (Nemo) belong to a group of fish that can be easily bred, the same, unfortunately, cannot be said of surgeonfish (Dory), who once fished out of the reefs leave a gap in the ecosystem there.

Fishing with cyanide compounds, of course, not only harms the fish that are caught but also the corals and other reef dwellers. Coral bleaching occurs, as the toxin has harmful effects on the algal symbionts of corals. People who still wish to pursue their hobby and are concerned about marine conservation would be well advised to be thorough in obtaining information about the origins of their fish and to only buy species from trusted breeders.

The use of cyanide or dynamite for fishing is banned in most countries, but there is hardly any policing and illegal fishers are seldom caught in the act, so for people with meager means, the fat profits are all too tempting despite the health risks, the danger of accidents, or fines. It is, however, at the expense of this unique ecosystem and, in the long run, the basis of existence for us humans whose survival depends on healthy and productive reefs.

It isn't only edible and ornamental fish that provide attractive incomes. Trade in other creatures is also lucrative: sea cucumbers (see plate 16) are considered a delicacy in China and as such have all but disappeared from many of their natural habitats. Roughly 30,000 tons of the easily gathered sea cucumbers are caught annually, mostly in Southeastern Asia, the Caribbean, and the Red Sea. In Vietnam, for instance, in every supermarket, you can find dried sea cucumbers on the shelf next to sealed bags of dried seahorses, starfish, and dried shark fins. You could find these products at every stall at the market on Chàm Island (the biosphere reserve where we ran a training course). The dried and neutral-tasting sea cucumbers are considered an aphrodisiac in China and are also believed to cure high blood pressure, cancer, and dementia. Demand is high and the price so good that illegal fishing worldwide is booming. In many places the stocks are exhausted. Should these tubular creatures disappear from the ecosystem, they will leave huge gaps. Sea cucumbers are particularly important in tropical coral reefs low in nutrients because they recycle organic material on the seafloor, excreting them, in part, as distinct particles. Their excrement is a valuable fertilizer for reef organisms, as it contains the components they require for their development; for instance, the formation of their calcite skeletons.

Hungry for More

I can still remember when Fridays were fish day and Atlantic salmon (*Salmo salar*) was only served on holidays. Now you can find a whole range of packed salmon in the refrigerated section of every supermarket and, for many people, it's not just for special occasions. Canned tuna, sushi bars, and, if we feel like it, fresh or deep-frozen fish from all over the world are available to us any day of the week. This, of course, is all at the expense of global fish stocks, the creatures' well-being, and the environment. The fishing industry is responding to satisfy the ever-increasing demand for fish and fish products (such as fish oil supplements or fish meal for aquarium fodder).

Using high-tech fishing practices, shoals of fish are tracked down and mercilessly netted—even at depths of 7,000 feet. Our demand for seafood delicacies has meant that 33 percent of global commercially viable fish stocks are overfished, 60 percent have been fished to their maximum biological limits, and only 7 percent of stocks could theoretically cope with an increased catch quota. "Overfishing" means that there has been an excessive depletion of fish stocks, whereby more fish have been caught than can accumulate through natural reproduction or migration. According to the Food and Agriculture Organization (FAO), a specialized agency of the UN, in 1961, 18.71 million tons of fish were caught worldwide. In 1995 that figure had climbed to 92.36 million tons of fish. Since then, the catches recorded have stagnated but stayed at a roughly similar level (90.91 million tons in 2016)—a sure sign that the sea can no longer keep pace with the heavy pressure of the fishing industry.

When you travel to a Mediterranean country, grilled sardines, octopus, tuna, etc.—preferably with a picturesque harbor in the background and a bottle of good local wine on the table—all make the vacation memorable. Mediterranean cuisine is supposed to be healthy, and fish, shellfish, olive oil, and fresh vegetables and fruits have all been prepared and served there, according to respective national variations, since time immemorial. But even though the menus in these restaurants don't give the impression that fish stocks are dwindling, around 93 percent of fish stocks in the Mediterranean are overfished. In the last 50 years alone, 41 percent of marine mammals and 34 percent of all other fish have vanished from the Mediterranean due to overfishing, pollution, and climate change. The majority of the Mediterranean fishing fleets are small, composed of boats no bigger than 30 feet, landing about a quarter of the total catches. The remaining fish are taken from the sea, mostly illegally, by large trawlers. In 2016, in an attempt to stop the further loss of species diversity—and, not to mention, to sustain the livelihoods of the more than 300,000 people dependent on fish and fishing—the EU's European Commission passed a 10-year commitment to increase stocks. In the MedFish4Ever declaration, representatives of the thirteen northern and southern countries bordering the Mediterranean agreed on a plan for the sustainable management of fish stocks and on ways of combating illegal fishing. If the implementation of this agreement is successful, there is hope that in the future we will be able to enjoy fresh fish with a clear conscience.

The EU has also stipulated annual fish quotas for countries fishing in the North Atlantic and North Sea. Environmentalists don't believe that the quotas go far enough, as stocks will not have reached sustainable levels by 2020. The quotas are based on scientific recommendations weighing up the interests of the fishing industry against conservation concerns. Unfortunately, even these recommendations are ignored often enough by EU fishery ministers, so there are, in general, still too many fish being caught.

Some fishing practices are worse than others. Different nets are used depending on species desired and the fish's preferred level in the water column. For instance, to catch herrings, cod, and halibuts, the mainly coastal fishing boats use sea gillnets with 50-foot mouths and lengths of up to 9 miles. There are also nets that can be set for the open waters of the pelagic zones and ones that scour the seafloor. Fish of a particular species or size can be targeted by selecting predetermined depths and net mesh size. Once fish have swum into the nets, their heads catch in the mesh. The more they struggle to free themselves, the more entangled they become. The use of trawl nets is highly contentious, as the catch quantities are very high. In addition to the targeted fish, every year an estimated 90,000 seabirds and up to 150 porpoises drown in gillnets alone in the Baltic. Pelagic or midwater otter trawling nets are also used to catch schooling fish such as herrings, mackerels, and sardines. These nets are funnel-shaped, narrowing to the cod end, and look a little like sacks; they are kept open by floats and weights. And the open mouths of trawl nets can have a volume of some 250,000 square feet—enough room for 12 jumbo jets. The trawlers are very well equipped with high-tech devices like sonar and echo sounders for locating and catching schools of fish. These huge nets are dragged slowly by one or sometimes two trawlers, with the fish collecting at the cod end of the net. Bycatch is enormous in the various sea regions, and regularly dolphins, sharks, whales, and sea turtles die painful deaths in the nets. Besides the pelagic trawl nets, bottom trawl nets are used by ocean fisheries to catch flatfish, cod, hake, and shrimps living on or near the seafloor. These nets are even used at depths up to 6,500 feet. When they are dragged along the seafloor, they leave a trail of destruction behind them, obliterating sessile organisms like cold-water coral as well as other creatures living at the bottom of the ocean. Fisheries that use beam trawling practices also damage the seabed. This bag-like bottom trawl is attached to a heavy metal beam and dragged along the seafloor, gathering the fish and shrimps and leaving the bottom looking as though it had been plowed. This form of fishing doesn't differentiate between the wanted species and the unwanted ones living at the bottom of the sea. According to Greenpeace, for every pound of sole caught in the North Sea there are 6 pounds of bycatch. The World Wildlife Fund (WWF) reckons the figures are even higher, with estimates of 15 pounds of bycatch per pound of flatfish. Fishing with purse seine nets is somewhat more selective. These nets are laid in a circle round a school of fish and then drawn together. When used on the high seas, seine nets can be a bit more than a mile long and reach depths of 700 feet. The fish caught using purse seines—such as herring, mackerel, sardines, or tuna (oceanic bonitos)—are then pumped on board. The fisheries often work with fish aggregating devices (FADS), a floating platform intended to simulate protection and attract schools of fish. The fish seeking refuge from predators aggregate around the floating platforms, where purse seine nets are laid and pulled tight. Unfortunately, here too the bycatch is large, as species other than the targeted ones, such as sea turtles and dolphins, are also magically attracted by the FADS. Between the end of the fifties and beginning of the nineties, up to 7 million dolphins were trapped and killed as bycatch in the seine nets fishing for tunas. In the tropical Eastern Pacific, the yellowfin tunas (*Thunnus albacares*), in particular, are known to socialize with schools of spinner and spotted dolphin schools. The fishers know that, in all probability, below a school of dolphins are plenty of tunas and so lay the nets around the marine mammals in order to catch the coveted tunas. According to the Earth Island Institute, an environmental action group based in California, during that period the "worst mass slaughter of marine mammals in the history of mankind" occurred. Since then catching methods have improved and been made more "dolphin-safe," but this has done nothing for other creatures caught in the same nets, among them sharks and sea turtles.

People who don't want to do without tuna but value the sustainability of the catching methods make sure that the fish they buy is caught by fishing rod, as is the case in the Maldives. Pole- and line-caught tuna (of species not considered endangered) can also be bought in supermarkets. The pole and line method involves the angler standing at the stern of a boat, letting out a baited line, and reeling in the fish. If the fish are too small, then they are simply thrown back into the sea, thus ensuring that stocks are sustainable. The "dolphin-safe" or "dolphin-friendly" labels, on the other hand, sadly tell us nothing about other species caught in the nets or about the management of the individual stocks.

Another similarly controversial method is longline fishing. This involves a mainline, which can be as long as 80 miles, with more than 20,000 baited hooks attached at intervals by means of branch lines. This non-selective catching method is intended to catch sharks, mahi-mahi, tuna, cod, swordfish, and other valuable edible fishes. Longlines are lain horizontally between the seafloor and surface, and held in position with floats and weights. There are two kinds of longline fishing: demersal longlines, drawn along the seafloor at depths between 1,500 and 8,000 feet, and pelagic longlines, which hang nearer the surface. The extent of bycatch—seabirds, sea turtles, sharks, and rays—is high in both methods. According to the WWF, at present global fishing practices take into account bycatches of around 38.5 million tons—that is 40 percent of total global catches. Dead and half-dead creatures are treated like garbage and just discarded in the sea.

More Culture for Aquaculture

So, what do you do if you don't want to go without fish? Is aquaculture the solution? For a long time now, the rapid increase in fish consumption has not been met by catches in the wild alone. According to the FAO, the worldwide total fish production, both wild and farmed, will rise from 171 million tons in 2016 to 201 million tons by 2030. In the meantime, 50 percent of the demand for shellfish, freshwater fish, and saltwater fish have to be farmed in aquaculture to satisfy our enormous appetite for fish and fish products. But this isn't preventing the overfishing of the seas. If anything, it is encouraging it, as tunas, salmon, and other popular edible farmed fish are often predators and have to be fed enriched soybeans, processed offal, or fish caught in the wild. Wild mackerel or other small fish are processed to make fish oil or meal, pressed into pellets, and fed to the cultured fish. For each pound of farmed salmon, 8 pounds of fish food are needed—every single day. Farmed tunas need five times the amount, per pound, of fish food, so some 40 pounds of wild-caught fish are needed. The use of soybeans, which also constitute a large part of the feed, is also questionable. For the cultivation of soybeans, depending on the country of origin, rain forests are cleared and unique habitats are replaced by monocultures. A further problem stems from transportation. When fish oil, soybeans, and fish meal are combined with oxygen in freight space, it becomes flammable, so antioxidative substances are added to prevent explosions and to stop the fodder from rotting. According to a random sample study by Greenpeace, ethoxyquin, an antioxidant food preservative, was found in 38 of 54 analyzed samples. These sample were packed fish products from conventional aquaculture. Since ethoxyquin is suspected of having carcinogenic effects and being genetically harmful, environmental agencies advise buying fish from either sustainable wild catches or organic aquaculture.

Conventional aquaculture conceals other ecological problems. Because the factory-farmed animals are under enormous stress, they are more susceptible to diseases. This is why cultured animals are fed antibiotics and other medications mixed into their fodder. These drugs, together with excrement and food remains, fall through the floating netted enclosures into rivers and the sea. The food remains spiked with antibiotics and over time accumulates in the sediment on the seafloor. In fact, more than 65 percent of fish fodder isn't eaten but sinks to the bottom. Not only is this harmful to the creatures living there but it also affects human beings. A Chinese study conducted in 2017 showed that the detritus (food remains) contained bacteria-resistant genes that in turn made microbes immune to antibiotics. This is bad news. If multi-resistant germs proliferate in the marine environment, they could eventually enter the food chain. Jing Wan and his research team at the Dalian Maritime University analyzed five commercial fish meal products and discovered over 1,300 different antibiotic-resistant genes. The researchers recommended targeted testing of animal meal for antibiotic residues and antibiotic-resistant genes in the future to prevent further spreading of multi-resistance germs. Suitable feeding strategies or efficient microbial substances also need to be developed in order to eliminate the fish meal residues already in the environment.

Another problem, especially affecting Norwegian fish farms, are parasitic attacks on cultured fish. One of Europe's most popular edible fish, the salmon, is prone to sea lice infestations. The salmon louse (*Lepeophtheirus salmonis*) is a tiny crab that attaches itself to the salmon, preferably somewhere in the head region, and then burrows its way deeper into the flesh. The parasites live off the mucus, skin, and blood of the salmon, which are then weakened and more susceptible to disease. As large numbers of salmon are kept in cramped netted enclosures, the salmon lice can spread rapidly, rather like head lice in nursery schools. In attempts to control these parasitic outbreaks, some producers treat the fish with pesticides; others try introducing cleaner fish into the enclosures in the hope of finding a biological solution; and others still use a combination of both practices. Pesticides that are spread farther afield by sea currents frequently cause environmental problems. Although there is no conclusive evidence, they are suspected to be the cause of the mass mortality of crustaceans in Norwegian fjords. If the salmon louse infestation among farmed fish is severe, then stocks have to be culled. As the larvae of salmon lice are distributed on currents, the lice also infect other fish and wild salmon outside the enclosures. These fish cannot be treated with chemicals or cleaner fish. According to the Norwegian Seafood Council, in 2016 alone, 53 million fish died from infestations.

A further problem with aquaculture is the destruction of marine habitats. In Vietnam and other Asian countries and in Central America, entire mangrove forests have been cleared for shrimp farming. But mangroves are very important ecosystems. They protect coastlines from storm flooding, tsunamis, and erosion while providing a vital habitat for specialized organisms and a nursery for a diversity of creatures that later populate coastal ecosystems, such as coral reefs. In addition to rain forests and coral reefs, mangrove forests are some of the most productive ecosystems on Earth. Once the mangrove forests disappear, then many species from coastal ecosystems also disappear, which also affects coastal fisheries. Catches decrease in precisely the locations that have been cleared most extensively. In 2018, scientists Thuy Dang Truong and Luat Huu Do discovered that in Vietnam, for instance, the density of mangroves has decreased drastically in recent decades. Since 1995, some mangrove forests in southern Vietnam have been allotted to households for clearing and cultivation; within the regulations, they are permitted to use 20 to 40 percent of their share of the mangrove forest for cultivating shrimps. Most households maintain a mixed shrimp and mangrove cultivation system where shrimp ponds are alternated with patches of mangroves. However, the farmers often use larger areas than those allowed for shrimp farming, the consequence being that mangroves are excessively cleared. With the use of chemical pesticides and antibiotics, the shrimp farms are so polluted after a few years that reforestation of the mangroves is virtually impossible. Within the last 50 years alone, 35 to 50 percent of all mangrove forests worldwide have been destroyed, with shrimp farming playing a considerable role. It isn't just the loss of these unique ecosystems and the related disappearance of species that is worrying. It's also the loss of mangroves as an efficient carbon reservoir. The CO_2 absorbed from the atmosphere by plants is transformed via photosynthesis to biomass—for every 2.5 acres 159 tons of pure CO_2 are absorbed. But most of the CO_2 remains in the dark sludge of the mangroves—on average an incredible 800 tons per 2.5 acres. Altogether, mangrove forests worldwide store between 4 and 20 billion tons of CO_2. This makes them important aids against climate change. If the mangrove forests are destroyed, the stored CO_2 reenters the atmosphere—all for the sake of some cheap shrimps.

225

The dumping prices of the seafood industry harm the environment and, ultimately, consumers. Unfortunately for people—particularly in the Western world—the deep-frozen salmon filets or the beautifully arranged shrimp nigiri on their plates in a sushi bar don't come with labels warning of the destruction of mangroves, or describing the suffering of animals, or drawing attention to the modern slavery on the fishing boats. To satisfy the demand for fish and shellfish, there have to be better solutions.

THE FUTURE
OF THE SEA = OUR
FUTURE

~~~~~~

THE WELFARE OF the "Blue Wonder" lies in our hands: First, as consumers, we need to make smarter purchasing decisions in a wide range of areas that will contribute to a recovery of the sea and sea life. Second, more intensive research into the effects of industry and pollution on the deep sea is needed to prevent the destruction of habitats out of greed. Last but not least, we need groundbreaking political rules and laws that will protect the sea.

Naturally, I enjoy eating fish and shellfish on occasion, and I often ask myself whether the information on the packaging holds true. Deciding what to buy from the refrigerated area of supermarkets has become even more difficult after research, in September 2018, by the ARD (a respected German broadcaster) led to heavy criticism of the Marine Stewardship Council (MSC) certification. The blue fish label on more than 50 percent of all fish products sold in Germany traditionally stood for sustainable wild seafood. Producers could use the symbol when they fulfilled three standards of the MSC: sustainable fish stocks, effective fisheries management, and healthy marine ecosystems. So far so good, but scientific research tells another story. Apparently, fisheries practicing bottom trawling—thus turning the seafloor into a desert landscape— were being granted certification, as bottom trawling was accepted by the MSC. Tuna fisheries in Mexico were also given certificates—thus accepting the killing of dolphins as bycatch in the purse seines (however, according to the MSC, it was no more than 500 per year). The findings of the ARD research suggested many times that figure. Scientists and environmentalists then demanded radical improvements and more transparency in the certification process.

Greenpeace advises that we not rely on certification alone, but to look at the fishing methods and catching areas. From experience, I know this can be pretty complicated. A number of apps that can be downloaded onto smartphones are available to offer advice. The fish advisor app from Greenpeace, for example, gives recommendations based on fishing methods and areas as well as sustainability. But even though this advice is helpful, a lot of responsibility still lies with the consumer. Not everyone has the capacity, energy, and time to check out all the various certifications. In 2018, the Consumer Advice Agency in Hamburg noted that a national sustainability seal with consistent criteria would make a lot of sense and is long overdue.

People who want to be absolutely sure have to either stop eating fish altogether or buy targeted products from organic aquaculture. Certification from Naturland guarantees that the keeping of animals is in accord with certain predefined standards and that the waters and surrounding ecosystems are undisturbed, that no chemicals or biogenetics are used, and that the fodder conforms to ecological norms as well as supports the high standard of social treatment of people working and living on their approved operations.

As demands for fish cannot be satisfied without aquaculture, and as this form of fish farming is allowing stocks of wild fish to recuperate, we need more innovative approaches. A proper sustainable project has been developed by a team from the University of Applied Sciences in Saarbrücken. The SEAWATER Cube is, according to the young team, a "biotechnological, complex aquaculture system for the failsafe and compatible production of marine fish species... the first small-scale, compact, and ready-for-use units, produced in small batches, allows [SEAWATER Cube] to provide regional and consumer-oriented production of fresh sea fish of the highest quality." Even though this project is in its early days, mobile, enclosed recirculation systems like the SEAWATER Cube do have the potential to produce top-quality saltwater fish, organically and sustainably and without long-distance transportation.

Once you have decided to give some thought to the subject and to be more conscious about your consumer habits, then the next problem arises: tackling the plastic problem (which in and of itself feels hopeless to many people). What can a single consumer do to combat the glut of plastic products in everyday life? Furthermore, reaching the right decision when shopping is anything but easy: Do I buy the organic bananas wrapped in plastic and the unsprayed lemons in the plastic net, or the loose fruit produced with pesticides? How do I know whether there are microplastics concealed in my shower gel, shampoo, or shaving foam, even if I manage to decipher the small print? Luckily, there is help to be found on the Internet. The smartphone app Beat the Microbead, for instance, can inform you whether a cosmetic product contains dangerous substances when you scan its barcode. On the Internet there are also plenty of lists of such products. People with good eyes or a magnifying glass in their pockets can read the ingredients themselves. A small hint: substances that contain microplastics often begin with "poly," "nylon," and "acrylate." In the meantime, there is a large range of alternatives to many products like shower gel, shampoo, and toothpaste that were previously available in mostly plastic exteriors and/or contained microplastics. So, there you are saving double on plastic. Whale calves and oysters, corals and sea turtles, will be thankful—and, ultimately, we humans, too.

Above all, government has to hold industry accountable. Instead of relying on dialogue and voluntary commitments, government, at long last, needs to enforce binding provisions to protect the environment from this surfeit of plastic waste. One of the first steps in the right direction took place on December 19, 2018. On that day, the negotiators of the EU parliament and the European states agreed to ban certain disposable plastic articles. It was an important first step, but you have to ask yourself why the ban only becomes fully effective starting 2021 when there are already plastic-free alternatives to tableware, straws, and disposable cups. Just to remind ourselves: every year 12.7 million tons of plastic end up in the sea, and it is believed that that this sum will double by 2030 and quadruple by 2050. Regardless, that something is finally being done at a policy-making level and that the changes are visible in day-to-day life is a positive step.

It was only about five years ago that majority of people who attended my lectures had never heard of microplastics. Today, the majority of my audiences know where and in which products these microplastics can be found. Stores selling products without any packaging whatsoever are shooting up like mushrooms, reusable cups made of sustainable materials can be found everywhere, and many supermarkets no longer supply plastic bags at the checkout.

Changes *are* taking place, which is great and encouraging. I am often asked where my motivation to tirelessly fight for healthy oceans comes from. Some days, when the bad news outweighs the good news, I ask myself the same question. When I walk along a beach gathering litter, cut through old fishing lines under water in coral reefs, or free dying marine creatures from nets, I burn with rage at the stupidity and ignorance of humans and think about just giving it all up. But then I remind myself of the progress: the days when I hold courses for instructors and workshops for trainees who, in turn, will spread their knowledge to their children and friends. The feedback that I get after such events reassures me that, collectively, many people want to and can change the way things are.

I don't live totally without plastic—to be honest, I still find the thought of doing so extremely difficult. The very few people who do manage to live plastic-free have my complete respect. However, if we are aware of what happens to the disposable articles we buy, where they end up, and the impact they have on humans as well as animals, we are in a better position to do without them the next time we go shopping. There are plenty of alternatives: bulk regional food products, cosmetics without microplastics, sustainable fish, resource-saving secondhand clothing, reef-saving sunscreens.

The fact that sun-protection products that don't harm coral reefs are now available shows a growing awareness that we need to protect the sea.

Of course, the seas will not be saved by these efforts alone. It will require larger and more far-reaching measures, particularly on global and political levels. A significant factor, if not the most important, is a radical reduction in the amount of plastic waste that we produce and that reaches the seas via all the rivers in the world but particularly the aforementioned top ten. This reduction in the amount of plastic waste reaching the sea can only be stanched by producing less plastic in the first place, which, as a consequence, will mean less waste is created. The ban on disposable plastic articles, which (only) comes into force in Germany in 2021, is a step in the right direction but is not nearly enough. A ban on microplastics in cosmetic products has long been implemented in the United States and Great Britain. Germany still relies on an appeal to the goodwill of the producers while, as Thomas Mani's research team already published in 2015, at least 192 million microplastic particles flow into the North Sea via the Rhine every day. This voluntary commitment is kowtowing to industry and is highly unlikely to solve the problem of environmental pollution by microplastics. A few strict laws on the use of microplastics and disposable plastics in general, as well as effective processing of wastewater, would lead to an improvement in the situation.

Once plastic has reached the sea it is almost impossible to remove it. Getting rid of the plastic waste that has accumulated in the gyres of the "Big Five" would certainly be a huge advance, but it has so far proven impossible. The Ocean Cleanup project, using a sort of floating comb-like device in which plastic is trapped and gathered, is trying to address the problem, but current technology is not yet sophisticated enough. Whether it will succeed is the subject of hot debate—there are concerns over the dangers of bycatch and the destruction of marine habitats, including that of the pleustons (the organisms living on or near the water–air interface). Still, it is thanks to projects like Ocean Cleanup that recently there has been more media interest in the plastic problem (even though the pollution of the seas by microplastics had been flagged in scientific publications in the early seventies). Since most of the plastic doesn't float on the surface but disappears to the deep-sea zones, it is difficult to imagine ever ridding the oceans of all plastic. This is why we must ensure that less plastic is made and less flows into the oceans.

We can tackle this together by consuming more consciously and by putting pressure on industry and government to secure our seas so they are healthy and full of life for subsequent generations. Or to put it in the words of the marine biologist Sylvia Earle: "We need to respect the oceans and take care of them as if our lives depended on it. Because they do."

ACKNOWLEDGMENTS

〰️

THE BLUE WONDER would never have materialized had I not received an email one day in which I was asked whether I would consider writing a book about the sea. This email was sent by my brilliant literary agent, Alfio Furnari, from the Landwehr & Cie agency. He deserves my special thanks, as he was the one who emboldened me to undertake this adventure. Dear Mr. Furnari, your shot in the dark hit the mark!

Before I started writing *The Blue Wonder* I had only dreamed of making the marvels of the ocean accessible to a worldwide audience. With the fantastic, charismatic, and funny Rob Sanders from Greystone Books, this dream became reality. Dear Rob, I cannot thank you enough for believing in me! A big thank-you also goes to the rest of the Greystone family, who supported me so dearly.

I'd like to think that my English isn't too bad, but it certainly isn't good enough to translate a whole book. This was done by the talented Jamie McIntosh, with editing by Tracy Bordian. Dear Jamie, dear Tracy, it was such a pleasure working with you. Many thanks for your patience and dedication to making this book available to English readers! Sometimes words and meanings get lost in translation and without proofreading professionals such as Alison Strobel the translated versions would not be as good as the original. Thank you very much, Alison, for putting the finishing touches on the English version!

A special thank-you goes to Jill Heinerth for providing such a lovely foreword to the English edition. I feel very honored that my book was able to impress and touch such a brave, experienced, adventurous, and highly professional woman and fellow lover of the blue wonder.

The person who ensured that my book was originally published by my favorite German publisher is the wonderful Jessica Hein, my editor at Ludwig Verlag. Many thanks, Ms. Hein, for believing in my book and making it possible for me to reach a wider public and share about a habitat that shapes all of our lives. I thank you for all of your efforts and enthusiasm. A big thank-you also goes to the whole team at Ludwig Verlag for the incredibly warm reception that awaited me in Munich. Your belief in my book makes me proud!

Writing doesn't always just flow; sometimes composing sentences becomes cumbersome and motivation slumps. My fantastic editor Dr. Angelika Winnen was the oil in the cogs of my book and the feather that tickled out my joy of writing. Dear Dr. Winnen, I thank you from the bottom of my heart for your support, words of encouragement, and what were quite likely nerve-racking finishing touches.

The paths we take are not always strewn with ideas, and inspiration can sometimes be harder to find than last year's Easter eggs. My friends and colleagues were the most resourceful egg hunters and provided me with even more questions and challenges: Dr. Hanna Schuster, who mulled over the craziest scientific queries and would not give up until she found the answers; Laura Riavitz, who continued to impress with her knowledge of corals—I couldn't have imagined a better or more competent colleague for our project in Vietnam; Jeannine Fischer, without whom the chapter about the kooky, funny, and completely bizarre sexual practices of some species would never have happened; Dr. Lauren Hall, for the Skype calls that were always just the thing to distract me from my work; and last but not least, Dr. Daniel Pauly, for his valuable input. Dear Daniel, it's such an honor to learn from the best—thank you!

Sincere thank-yous also go out to Barbara Matheis, with whom I share heartfelt laughter; Angela and Nils Jensen, who waded through all the paperwork and finally made our project *The Blue Mind* official; Dr. Chrysoula Gubili and Dr. Sokratis Loucaides, who supplied publications that I couldn't access; Alexandra Christian, for the long walks with the dogs when I could clear my head; Ms. Funk, for precious relaxation techniques; and Sarah Bels, for shelter in Barcelona where I was seeking and found inspiration. Janina D'Agostino, Christina Rosenthal, Nicola Schönenbach, Julia Lang, Nadine Jansen, Heike Klees, Danja Mansour, Dr. Sebastian Meller, Prof. Dr. Uwe Waller, Catherine Mentz, Katrin Eckart, Sarah Westphal, and my godmother, Angelika Frisch—thank you for your belief in my book project. Angela Jensen, Tam Sawers, David Molina Ferrer, and Pierre Bouras, thanks for the beautiful photos.

My parents and brothers never tired of motivating and spurring me on. A huge thank-you goes to my brother Fabian for the almost daily telephone calls that enabled me to let off steam. Dear Fabian, thank you for making me laugh so often! To my mother, thank you for having a sympathetic ear and for your belief in me. To my father, thanks for the valuable information on plastics. He spent almost all of his working life with plastics, and I couldn't have found a better-qualified contact for my inquiries on the subject. Dear Ma, dear Pa, without you I could never have achieved the job of my dreams—thank you. A big thank-you to my brother Tobias and his partner Verena, who cared for Oskar, enabling me to travel with a clear mind. Dear Tobias, dear Verena, it is good to know that I can always rely on you.

Cliché or not, I must credit coffee, red wine, and chocolate, without which the inevitable bouts of writer's block would have been unbearable.

SOURCES AND FURTHER INFORMATION

The Secret Global Domination of Plankton

Albert, D. J. (2011): What's on the mind of a jellyfish? A review of behavioral observations on *Aurelia* sp. jellyfish. *Neuroscience and Biobehavioral Reviews* 35 (3): 474–482.

Boyce, D. G. et al. (2010): Global phytoplankton decline over the past century. *Nature* 466: 591–596.

Haeckel, E. H. (2016): *Kunstformen der Natur.* (Art Forms in Nature.) Wiesbaden: Marixverlag.

Houghton, I. A. et al. (2018): Vertically migrating swimmers generate aggregation-scale eddies in a stratified column. *Nature* 556: 497–500.

Lana, A. et al. (2011): An updated climatology of surface dimethlysulfide concentrations and emission fluxes in the global ocean. *Global Biogeochemical Cycles* 25 (1): GBI004.

Last, K. S. et al. (2016): Moonlight drives ocean-scale mass vertical migration of zooplankton during the Arctic winter. *Current Biology* 26: 244–251.

Piraino, S. *et al.* (1996): Reversing the life cycle: Medusae transforming into polyps and cell transdifferentiation in *Turritopsis nutricula* (Cnidaria, Hydrozoa). *The Biological Bulletin* 190 (3): 302–312.

Sacova, M. S. & Nevitt, G. A. (2014): Evidence that dimethyl sulfide facilitates a tritrophic mutualism between marine primary producers and top predators. PNAS 111 (11): 4157–4161.

Sardet, C. (2015): *Plankton: Wonders of the Drifting World.* Chicago: University of Chicago Press.

Savoca M. S. *et al.* (2016): Marine plastic debris emits a keystone infochemical for olfactory foraging seabirds. *Science Advances* 2 (11): e1600395.

Sekerci, Y. & Petrovskii, S. (2018): Global warming can lead to depletion of oxygen by disrupting phytoplankton photosynthesis: A mathematical modelling approach. *Geosciences* 8 (6): 201.

Uye, S. (2008): Blooms of the giant jellyfish *Nemopilema nomurai*: A threat to the fisheries sustainability of the East Asian marginal seas. *Plankton and Benthos Research* 3 (Suppl.): 125–131.

WEBSITES

CNN: www.edition.cnn.com/2014/08/28/world/asia/canimmortaljelly fishunlockeverlastinglife/index.html?hpt=hp_c4. Last accessed on 4 January 2019.

Spiegel Wissenschaft (Science Mirror): www.spiegel.de/wissenschaft/natur/ klimaphaenomene-wiealgen-wetter-machen-a-756865.html. Last accessed on 4 January 2019.

Coral Reefs—The Nurseries of the Sea

Balcombe, J. (2016): *What a Fish Knows: The Inner Lives of Our Underwater Cousins.* New York: Scientific American/Farrar, Straus and Giroux.

Beattie, M. *et al.* (2017): The roar of the lionfishes *Pterois volitans* and *Pterois miles. Journal of Fish Biology* 90 (6): 2488–2495.

Becker, J. H. A. et al. (2005): Cleaner shrimp use a rocking dance to advertise cleaning service to clients. *Current Biology* 15 (8): 760–764.

Bshary, R. et al. (2006): Interspecific communicative and coordinated hunting between groupers and giant moray eels in the Red Sea. *PLoS Biology* 4 (12): e431.

Erisman, B. E. & Rowell, T. J. (2017): A sound worth saving: Acoustic characteristics of a massive fish spawning aggregation. *Biology Letters* 13 (12): 20170656.

Eyal, G. et al. (2015): Spectral diversity and regulation of coral fluorescence in a mesophotic reef habitat in the red sea. *PLoS ONE* 10 (6): e0128697.

Fine, M. & Parmentier, E. (2015): "Mechanisms of Fish Sound Production." In: Ladich, F. (ed.): *Sound Communication in Fishes: Animal Signals and Communication*, Vol. 4. Vienna: Springer-Verlag, 77–126.

Patek, S. N. et al. (2004): Deadly strike mechanism of a mantis shrimp. *Nature* 428: 819–820.

Perdicaris, S. et al. (2013): Bioactive natural substances from marine sponges: New developments and prospects for future pharmaceuticals. *Natural Products Chemistry & Research* 1: 114.

Pinto, A. et al. (2011): Cleaner wrasses *Labroides dimidiatus* are more cooperative in the presence of an audience. *Current Biology*, 21 (13): 1140–1144.

Raihani, N. J. et al. (2011): Male cleaner wrasses adjust punishment of female partners according to the stakes. *Proceedings of the Royal Society B: Biological Sciences* 279 (1727): 365–370.

Smith, E. G. et al. (2017): Acclimatization of symbiotic corals to mesophotic light environments through wavelength transformation by fluorescent protein pigments. *Proceedings of the Royal Society B: Biological Sciences* 284 (1858), rspb.2017.0320.

Thoen, H. H. et al. (2014): A different color vision in mantis shrimp. *Science* 343 (6169): 411–413.

Vail, A. L. et al. (2013): Referential gestures in fish collaborative hunting. *Nature Communications* 4: 1765.

(In)Finite Blue

Caesar, L. *et al.* (2018): Observed fingerprint of a weakening Atlantic Ocean overturning circulation. *Nature* 556: 191–196.

Chapman, D. D. *et al.* (2007): Virgin birth in a hammerhead shark. *Biology Letters* 3 (4): 425–427.

Domenici, P. *et al.* (2014): How sailfish use their bills to capture schooling prey. *Proceedings of the Royal Society B: Biological Sciences* 281 (1784).

Leigh, S. C. *et al.* (2018): Seagrass digestion by a notorious "carnivore." *Proceedings of the Royal Society B: Biological Sciences* 285 (1886).

Mansfield, K. L. *et al.* (2014): First satellite tracks of neonate sea turtles redefine the "lost years" oceanic niche. *Proceedings of the Royal Society B: Biological Sciences* 281 (1781).

Myers, R. A. *et al.* (2007): Cascading effects of the loss of apex predatory sharks from a coastal ocean. *Science* 315 (5820): 1846–1850.

Neukamm, M., ed. (2014): *Darwin heute. Evolution als Leitbild in den modernen Wissenschaften.* (Darwin Today: Evolution as a Model in Modern Sciences.) Darmstadt: WBG.

Thornalley, D. *et al.* (2018): Anomalously weak Labrador Sea convection and Atlantic overturning during the past 150 years. *Nature* 556: 227–230.

Worm, B. *et al.* (2013): Global catches, exploitation rates, and rebuilding options for sharks. *Marine Policy* 40: 194–204.

WEBSITE

Max-Planck-Institut für Meteorologie (Max Planck Institute for Meteorology): www.mpimet.mpg.de/kommunikation/fragen-zuklima-faq. Last accessed on 16 March 2019.

The Secrets of the Deep

Hartwell, A. M. *et al.* (2018): Clusters of deep-sea egg-brooding octopods associated with warm fluid discharge: An ill-fated fragment of a larger, discrete population? *Deep Sea Research Part I: Oceano-Graphic Research Papers*, 135: 1–8.

Lee, W. L. *et al.* (2012): An extraordinary new carnivorous sponge, *Chondrocladia lyra*, in the new subgenus Symmetrocladia (Demospongiae, Cladorhizidae), from off of northern California, USA. *Invertebrate Biology* 131 (4): 259–284.

Partridge, J. C. (2012): Sensory ecology: Giant eyes for giant predators? *Current Biology* 22 (8): R268–R270.

Purser, A. *et al.* (2016): Association of deep-sea incirrate octopods with manganese crusts and nodule fields in the Pacific Ocean. *Current Biology* 26 (24): R1268–R1269.

Pauly, D. (2021): The gill-oxygen limitation theory (GOLT) and its critics. *Science Advances* 7 (2): eabc6050.

Robison, B. H. *et al.* (2014): Deep-sea octopus (*Graneledone boreopacifica*) conducts the longest-known egg-brooding period of any animal. *PLoS One* 9 (7): e103437.

Robison, B. H. & Reisenbichler, K. R. (2008): *Macropinna microstoma* and the paradox of its tubular eyes. *Copeia* 4: 780–784.

WEBSITES

Discovering Hydrothermal Vents: www.whoi.edu/feature/history-hydrothermal-vents/discovery/1977.html. Last accessed on 4 January 2019.

IUCN: www.iucn.org/news/secretariat/201807/draft-miningregulations-insufficient-protect-deep-sea-%E2%80%93-iucnreport. Last accessed on 31 January 2019.

National Ocean Service: https://oceanservice.noaa.gov/facts/pressure.html. Last accessed on 8 January 2020.

Sex and the Sea

Aarestrup, K. *et al.* (2009): Oceanic spawning migration of the European eel (*Anguilla anguilla*). *Science*, 325 (5948): 1660.

Harris, H. S. *et al.* (2010): Lesions and behavior associated with forced copulation of juvenile Pacific harbor seals (*Phoca vitulina richardsi*) by southern sea otters (*Enhydra lutris nereis*). *Aquatic Mammals* 36 (4): 331–341.

Kawase, H. *et al.* (2013): Role of huge geometric circular structures in the reproduction of a marine pufferfish. *Scientific Reports* 3: 2106.

Ramm, S. A. *et al.* (2015): Hypodermic self-insemination as a reproductive assurance strategy. *Proceedings of the Royal Society B: Biological Sciences,* 282 (1811).

Russell, D. G. *et al.* (2012): Dr. George Murray Levick (1876–1956): Unpublished notes on the sexual habits of the Adélie penguin. *Polar Record* 48 (4): 387–393.

WEBSITE

Video of the mating rites of deep-sea anglerfish *Caulophryne jordani*: www.youtube.com/watch?v=anD1lMvgNwk. Last accessed on 6 February 2019.

Endangered Blue Wonder

Abts, G. (2016): *Kunststoff-Wissen für Einsteiger.* (Plastic Knowledge for Beginners.) München: Carl Hanser Verlag.

Allen, A. S. *et al.* (2017): Chemoreception drives plastic consumption in a hard coral. *Marine Pollution Bulletin* 124 (1): 198–205.

Chiba, S. *et al.* (2018): Human footprint in the abyss: 30-year records of deep-sea plastic debris. *Marine Policy* 96: 204–212.

Germanov, E. S. *et al.* (2018): Microplastics: No small problem for filter-feeding megafauna. *Trends in Ecology & Evolution* 33 (4): 227–232.

Han, Y. *et al.* (2017): Fish meal application induces antibiotic-resistant gene propagation in mariculture sediment. *Environmental Science & Technology* 51 (18): 10850–10860.

Hughes, T. P. *et al.* (2018): Spatial and temporal patterns of mass bleaching of corals in the Anthropocene. *Science* 359 (6371): 80–83.

Jambeck, J. R. *et al.* (2015): Plastic waste inputs from land into the ocean. *Science* 347 (6223): 768–771.

Kaza, S. *et al.* (2018): *What a Waste 2.0: A Global Snapshot on Solid Waste Management to 2050.* Urban Development Series. Washington: World Bank Group.

Lamb, J. B. *et al.* (2018): Plastic waste associated with disease on coral reefs. *Science* 359 (6374): 460–462.

Mani, T. *et al.* (2015): Microplastics profile along the Rhine River. *Scientific Reports* 5: 17988.

Rochman, C. M. *et al.* (2015): Anthropogenic debris in seafood: Plastic debris and fibres from textiles in fish and bivalve sold for human consumption. *Scientific Reports* 5: 14340.

Schmidt, C. *et al.* (2017): Export of plastic debris by rivers into the sea. *Environmental Science & Technology* 51 (21): 12246–12253.

Truong, T. D. & Do, L. H. (2018): Mangrove forests and aquaculture in the Mekong River Delta. *Land Use Policy* 73: 20–28.

Wilcox, C. *et al.* (2018): A quantitative analysis linking sea turtle mortality and plastic debris ingestion. *Scientific Reports* 8: 12536.

WEBSITES

Climate protection report of the German government 2017: www.bmu.de/publikation/klimaschutzbericht-2017. Last accessed on 16 March 2019.

Climate protection report of the German government 2018: bmu.de/fileadmin/Daten_BMU/Download_PDF/Klimaschutz/klimaschutz bericht_2018_bf.pdf. Last accessed on 22 February 2019.

Food and Drug Administration (U.S.): www.fda.gov/food/food-additives-petitions/bisphenol-bpa-use-food-contact-application.

Greenpeace: www.greenpeace.de/themen/meere/app-fuer-nachhaltigen fisch. Last accessed on 16 March 2019.

IEA: www.iea.org. Last accessed on 16 March 2019.

Live Science: www.livescience.com/63597-compost-trash-in-landfills.html.

Ministerial Conference on the Sustainability of Mediterranean Fisheries (paper): www.ec.europa.eu/fisheries/sites/fisheries/files/2017-03-30-declaration-malta.pdf. Last accessed on 21 February 2019.

Plastics Europe: www.plasticseurope.org/de. Last accessed on 16 March 2019.

Plastic Pollution Coalition: www.plasticpollutioncoalition.org/blog/2019/3/6/157000-shipping-containers-of-us-plastic-waste-exported-to-countries-with-poor-waste-management-in-2018.

WDR: www1.wdr.de/mediathek/video/sendungen/die-story/video-dasgeschaeft-mit-dem-fischsiegel-die-dunkle-seite-desmsc-100.html. Last accessed on 22 February 2019.

General Resources

Bundesinstitut für Risikobewertung (Federal Institute for Risk Assessment): www.bfr.bund.de/en.

FishBase: www.fishbase.org.

Food and Agriculture Organization of the United Nations: www.fao.org.

Greenpeace: www.greenpeace.org.

Hempel, G. *et al.* (2017): *Faszination Meeresforschung.* (Fascination With Marine Research.) Berlin: Springer-Verlag.

IPCC: www.ipcc.ch.

NABU: www.en.nabu.de.

NASA: www.nasa.gov.

National Oceanic and Atmospheric Administration: www.noaa.gov.

Ocean Quest: www.oceanquest.global.

SeaLifeBase: www.sealifebase.ca.

Umwelt Bundesamt (Environment Federal Office): www.umweltbundesamt.de/en.

UN Environment Programme: www.unenvironment.org.

World Ocean Review: www.worldoceanreview.com.

PHOTO INSERT CREDITS

~~~~~~~~~~

Plates 1, 14, 16, 17, 22, 23 (top): Copyright © Pierre Bouras

Plates 2, 3, 4, 5, 6, 12 (bottom), 13, 15, 18, 19, 20, 21, 23 (bottom), 24: Copyright © Frauke Bagusche

Plates 7, 11 (top): Copyright © Laura Riavitz

Plates 8, 9: Copyright © Tam Sawers

Plate 10: Copyright © Angela Jensen

Plates 11 (bottom), 12 (top): Copyright © David Molina Ferrer

INDEX

Note: "plate" followed by a number refers to the pages of the photo insert.

257

262